the iPhone app design manual

dave brown
with vicky roberts

the iPhone app design manual

create perfect designs for effortless coding and app store success

dave brown

with vicky roberts

ILEX

THE IPHONE APP DESIGN MANUAL

First published in the United Kingdom in 2014 by
ILEX
210 High Street
Lewes
East Sussex
BN7 2NS

Distributed worldwide (except North America)
by Thames & Hudson Ltd., 181A High Holborn,
London WC1V 7QX, United Kingdom

Publisher: Alastair Campbell
Creative Director: James Hollywell
Managing Editors: Nick Jones and Natalia Price-Cabrera
Commissioning Editor: Zara Larcombe
Art Director: Julie Weir
Editor: Ellie Wilson
Designer: Lisa McCormick

British Library Cataloguing-in-Publication Data
A catalogue record for this book is available from the British Library.

ISBN: 978-1-78157-140-8

Colour Origination by Ivy Press Reprographics

Printed and bound in China

10 9 8 7 6 5 4 3 2 1

For my mum and late father.

Thank you for always supporting me and for giving me the love and belief to succeed. I'll always love you more than I'll ever be able to say.

CONTENTS

INTRODUCTION

When I founded Apposing back in 2009, the app world was a much less congested space. The boom had not yet begun and the way that we interacted with technology was infinitely more stationary.

With a background in web design, I was well-versed in the processes and theories which led to a great website and experience for the user, but the advent of the mobile application posed a new area of interest. I became obsessed with the power that these apps held within their tiny frameworks. Carried in the pocket of the user wherever they went, apps allowed the mobile phone to become whatever it needed to be. You no longer needed to carry multiple digital, and in some cases physical, tools; they could all be neatly housed in one pocket-sized package.

Apps presented an exciting opportunity to connect with customers for brands and companies across the world, but more importantly, they offered designers a method of changing the digital landscape for the better. Age-old processes could be revisited; app designers had the opportunity to study the way we interact with things to see if they could be improved. Good app design was about more than taking the internet and making it accessible on your phone, it was an art form in its own right.

I founded Apposing with the notion of creating truly useful apps with the user in mind. I wanted to make valuable apps that generated excitement, pushed boundaries, and which solved problems that people might not have known that they had. A fledgling soldier on the digital frontline, Apposing was at that time a one-man band with a vision.

Fastforward to 2014, and I've achieved what I set out to do. We get to work with some of the world's biggest brands, creating apps which have evolved the way that customers shop, work, and play. We've managed to amass over one million downloads across Android and iOS. Now a team of thirteen, our ethos remains the same: we spend a lot of time researching who an app's users will be and the business that we're creating the app for so that we can create something that people will wonder how they coped without it. Though the app industry is by no means a failsafe way to make money, our company's approach has allowed us to make a living doing what we love most, and having a lot of fun in the process.

Now people well and truly understand the power of the app, with more making the move into app development than ever before. New and improved processors which arrive with each new model of iPhone give designers and developers new ways of exploiting the technology to create something game-changing. The opportunities are endless.

This book will take you through our method for building an iPhone app. From defining your app's primary function through to deciding on which screenshots to submit to the App Store, we offer practical and easy-to-follow advice gleaned from our experience in the business. Featuring our top tips for creating an app that's not only a joy to behold, but also has user-experience at its core, our jargon-free book explains how to make your idea shine.

We've also amassed a plethora of useful templates, resources, tutorials, and videos to make the transition to app designer a little easier, available here: ***www.iphoneappdesignmanual.com***

The app world is constantly changing. If you want to keep up to date with what's new and developments you can take advantage of, then this book alone won't be enough. While we can show you how to plan and execute your design, keeping up to date with the mobile sphere will serve you well. We've never stopped learning.

I hope that you enjoy finding new ways to change the digital landscape as much as we do. When we're brainstorming, no idea is too crazy; we explore each and every one. Some of our best ideas have been hatched this way.

Dave Brown @daveapposing
MD and Founder of Apposing

CHAPTER ONE

WHAT IS UNIQUE ABOUT DESIGNING MOBILE APPS?

The age of the smartphone is upon us, and with it has come a surge of applications, or apps, which have evolved the way that we interact with our surroundings, each other, and brands and businesses forever.

Gone are the days when a mobile phone was used primarily for calls. We use our phones whilst watching a favorite TV show, to check we've not missed a recipe's vital ingredient when out shopping, and to upload pictures to social media on a night out. There are also a wealth of apps that save us time and money, and that have altered the way that we perform tasks indefinitely.

Though they've only been with us a relatively short amount of time, it's difficult to imagine how we'd cope without them. There's even an official word, "Nomophobia," for the fear of leaving our phones at home. But how does a mobile app differ from a website? What is it about these tiny pieces of software that has entwined itself with our day to day lives so tightly? This chapter aims to find out.

Designing for Mobile Vs. Web

While app and web design clearly share some traits and principles, approaching an app like a traditional website for a smaller screen would be a rookie mistake. Though successful and profitable websites have long had their roots planted firmly in user-centric design, if you've had any experience in designing applications, you'll know that the mobile user journey is an entirely different ball game.

Equally, if you're viewing an app as an extension of a website, you're failing to realize its full potential. Apps offer designers more opportunity to harness their creativity than ever before. If you're committed to understanding how people will use your app, how it could add benefit to their lives, and to solving problems people never knew they had, then designing a useful and successful app is well within your reach.

While today's consumers expect a quality and high-end interface as standard for both platforms, ultimately, it's the usability and function of a website or app that determines its success or failure. To put it simply, if a user cannot figure out a way to complete a task in a relatively short amount of time, then they will probably leave empty-handed and be unlikely to return.

Understanding the way a consumer uses a website has never been more important, but the advent of the mobile age has presented designers with a fresh set of challenges and considerations. Web design has taught us much about the importance of the user journey, but the simple fact that the cell (mobile) phone is handheld and rarely out of eyeshot means much of what we've learned needs to be adapted.

Usability is key to determining the success of both a website or a mobile app. The user must be able to grasp and carry out the primary function in a short amount of time.

The rise of the cell phone has taken the principles of web design and turned them on their head.

Sitting Pretty

Though the internet has never slept, using it was generally confined to the home, office, or cafes that offered an internet connection. Usually seated and with both hands free, the web user explored the internet via a screen, a keyboard, and a mouse, discovering information before they left the home, office, or cafe via a wired or wireless connection.

Savvy designers understand the science behind the way consumers view a web page, where they look first, and where to place things to keep tasks simple. As most people would view the design sitting and facing a monitor, web designers could take a lot of things for granted. Menus with mouse-activated drop-downs became commonplace; menu bars were placed at the top of the screen for easy viewing and navigation. Flash was also used extensively throughout. Web controls such as radio buttons and drop-down menus evolved from the inputs used by computers, such as a mouse and keyboard. Cell phones don't have these inputs, so they need a new set of interface controls. Likewise, Flash isn't supported by iOS, so web designers need to rethink this practice. While people will browse a full website from the comfort of their desk, mobile access can happen anywhere, at any time. Whether standing or sitting, whether on the couch or walking through a busy airport, the sheer number of ways we use our cell phone introduces the designer to a new set of rules and design consequences.

How We Access the Web Via Mobile

With over 1.8 billion of the world's four billion cell phones now a smartphone, mobile saturation is at an all-time high. Increasingly, consumers are using their cell phones as their first port of call to access the internet, with mobile access predicted to overtake desktop access by 2015 (Microsoft Tag, 2013).

As mobile popularity continues on the upward trajectory, the majority of businesses now have a mobile strategy in place. Facebook creator Mark Zuckerburg went as far as to say that as of 2013, his social network is now a "mobile first, mobile best" company. At the very least, web designers are creating mobile or "responsive" websites. A mobile site is generally considered to be the first step in creating a mobile presence, and, with the exception of games and apps such as Instagram, few apps are created without a company or brand having a website in place first.

THIS ACCESS TAKES FOUR MAIN FORMS:

1 **Users access the desktop version of a website via the decidedly smaller phone screen:** Clunky and with long load times, this type of access eats up a user's data allowance and can provide an irritating and long-winded experience. Users must pinch to zoom in to the action, and buttons are designed for a mouse click, not a stubby finger.

2 **The mobile site/application:** The mobile application takes two main forms. The first and more prominent is the responsive website. Designed to make browsing easier and quicker, the mobile site displays the same content as the website with the design automatically modified to suit the device it's being used on. Updating the desktop site will result in the same changes on a mobile screen. HTML5 responsive sites cannot currently use all of the device's functions, but can use the camera, accelerometer, and GPS.

The second is the dedicated mobile web application. Optimized for mobile use, a dedicated mobile site is very similar to a desktop website, but generally omits many of the features for a faster experience, almost a "website-lite." Users can also download a shortcut button for their phone. These exist entirely online. Without an internet connection, it is not possible to interact with the application. Benefits of this type of app for the developer include being able to

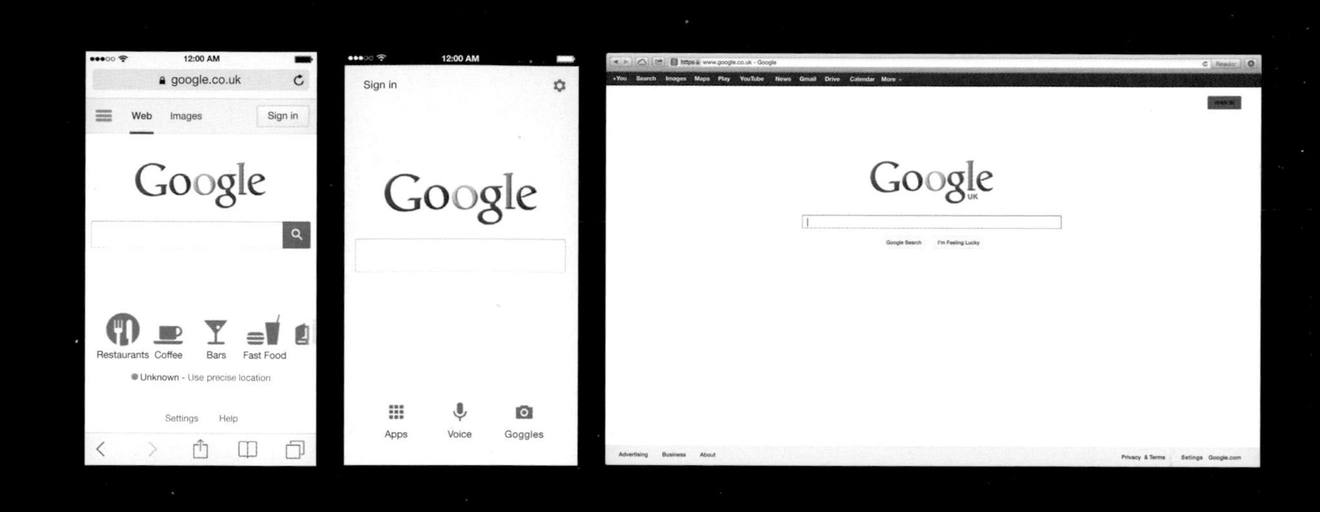

update the look, feel, and content whenever they feel the need. This type of app cannot use the phone's native functions, except for those built using HTML5, which can use the camera and GPS.

3 **Connecting through a native application:** Built specifically for the device, a native application runs physically on the phone it's installed on. This is the type we'll be dealing with in this book. Native applications can use the phone's functions, such as the camera and GPS, and most content and functions can be accessed without an internet connection. Native iOS apps are developed/written using the programming language Objective-C, while Android apps are developed/written in Java. To update a native app, users must download a new version through stores such as iTunes or Google Play. Unless the app has a built-in content management system, iOS developers must submit all changes to their app through the App Store and receive approval from Apple before it gets released.

4 **The hybrid application:** This type of application blurs the lines between web-based and native applications. Primarily it's a web-based app that uses programming language HTML5, but is wrapped in an Objective-C, native app shell (see chapter two for programming languages). This means that it can access the device's primary functions. This type of app is relatively new to the party and noticeably slower than its native counterpart, though developers aren't discounting its potential for dominance in the future.

How App Design Differs From Web Design

What are the key design differentiators between the two? As with web design, good mobile design has user experience at its core, so the consumer journey is just as important as the visual aspect. However, the way we consume mobile content is different. While a responsive web layout or web-based applications can offer users a static navigation, native applications offer a more interactive experience. The ability to pinch, swipe, and tap your way through a site instead of clicking or pushing buttons heavily influences the design and layout.

One Hand Can Be Better Than Two

As its very core, the word mobile suggests its users are just that. Applications that can be used single-handed offer benefits for consumers on the move. Allowing people to access their connecting train time while traversing a busy platform is preferable to a web-based app that requires both hands. The thumb is used almost exclusively during one-handed browsing, and designers must take this into consideration, keeping controls localized to the average thumb's reach. Placing buttons or menu bars at the top of an application as you would with many websites would mostly be a mistake in the world of app development, creating a jerky and strained journey for the user.

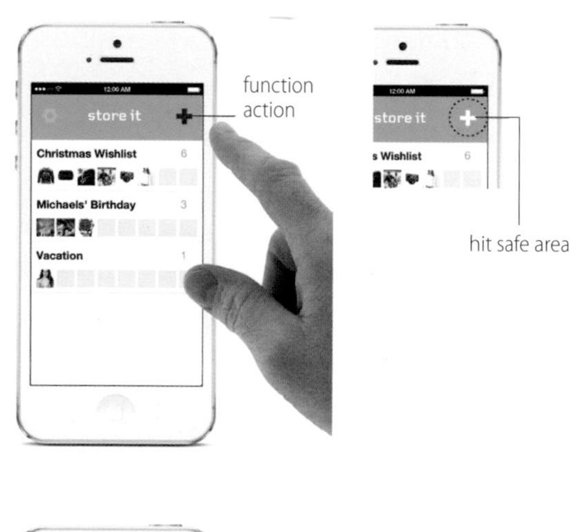

Similarly, functionality found too close to the device's edge can be difficult to tap precisely and should be avoided. Buttons also need to be of average fingertip or thumb size; Apple recommends a minimum of 88 pixels in height or width for the retina screen. Anything smaller becomes fiddly and can make function

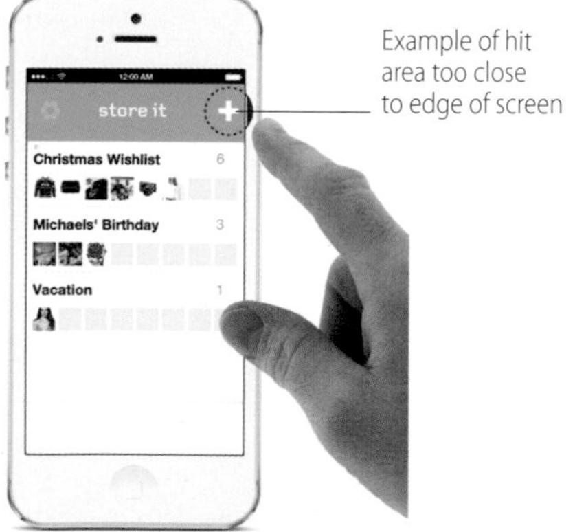

Example of hit area too close to edge of screen

accessibility a hassle for the consumer. Controls always need to be kept below where the action is, otherwise the user won't be able to see what they're up to onscreen.

Size Matters

When it comes to app design, size definitely matters. Browsing on the move, users don't want a hefty chunk of their data allowance eaten up by unnecessary images and functions. It's also worth noting that iOS apps upwards of 100MB can only be downloaded via a WI-FI connection—even if you've downloaded via your computer, the app will only appear on your phone when you're connected to WI-FI. Apps below the threshold can be downloaded over 3G connections; more desirable for the likes of travel-based apps you might not have known you'd need.

Less is More, but More is More

"Feature creep," adding new features to software to the detriment of the original concept or ease of use, is another concern for the app designer. An app overloaded with functions can be overwhelming for the user, more so than with website design as the real estate is infinitely smaller. Like websites, apps don't come with instruction manuals and the consumer must know almost intuitively how to use them. Designers must keep the context in which the app will be used in mind, at all times, ensuring that excess information is trimmed back. However, context is key. Restaurants that

host their full menu on their website, but decide to leave it out of their app for a streamlined experience are missing a trick. Through not fully exploring the context in which their users will utilize their app, they miss out a vital component in what their users need.

An app that sticks to its primary objective gets to the point and the information quickly, with minimum effort on the user's part, rendering itself the very definition of useful.

It's important to note that though some features are likely to be omitted from a mobile site, the extra phone functions on offer to the app designer could, and perhaps should, result in an app that does more than the mobile site. From personal scrapbooks in shopping apps to barcode scanners, great app designers integrate the phone's functionality into their design.

Typing is Tedious

Web designers have the luxury of end users with access to a full keyboard. Typing on a mobile is both time-consuming and trying; designers must ensure that any data collection requires the bare minimum of effort. A considerate designer would also ensure that the relevant keyboard (letters/numbers/symbols) appears for the task in hand. Users with an iPhone 5s or above enjoy the benefits of a fingerprint scanner. Housing up to five fingerprint scans on the phone, users can use their fingerprint to buy content or access functions without the need to type.

Explore the Extra Features

One of the exciting elements of app development is the wealth of features the phone offers. The camera, calendar, GPS tracking, and microphone can all add depth and further functionality to an idea in a way that can't be achieved with web design. The accelerometer, a sensor which uses tilts, swings, rotation, and motion gestures to control some aspect of the phone, adds an extra dimension to game controls in titles such as Doodle Jump. It can be employed to great effect replacing a button, with a great example of this being the "ignore call" function, in which iPhone users simply put their phone face down to forward the call to voicemail. Good app functionality is designed around these features, rather than using them as add-ons.

Unlike web designers, app designers have access to extra features that are part of the phone's hardware, such as the camera. Good apps make use of these effectively.

Art Imitates Life

The use of fingers, hands, and motion to control your app has led many designers to employ skeuomorphism—where a design feature, noise, or texture mimicking a real-world artifact (say the lined pages of a notepad or a camera's click) is used to make the functionality or look of an application more familiar or visually appealing. The technique can be employed to make the application easier to use: as it becomes more similar to real-world objects, how to use it becomes more obvious.

However, iOS 7 swerved dramatically away from the use of skeuomorphs, with a greater focus on clean and simple, two-dimensional graphics and interfaces. Some skeuomorphism remains, in apps such as iBooks, which still features wooden shelves that books are rested on.

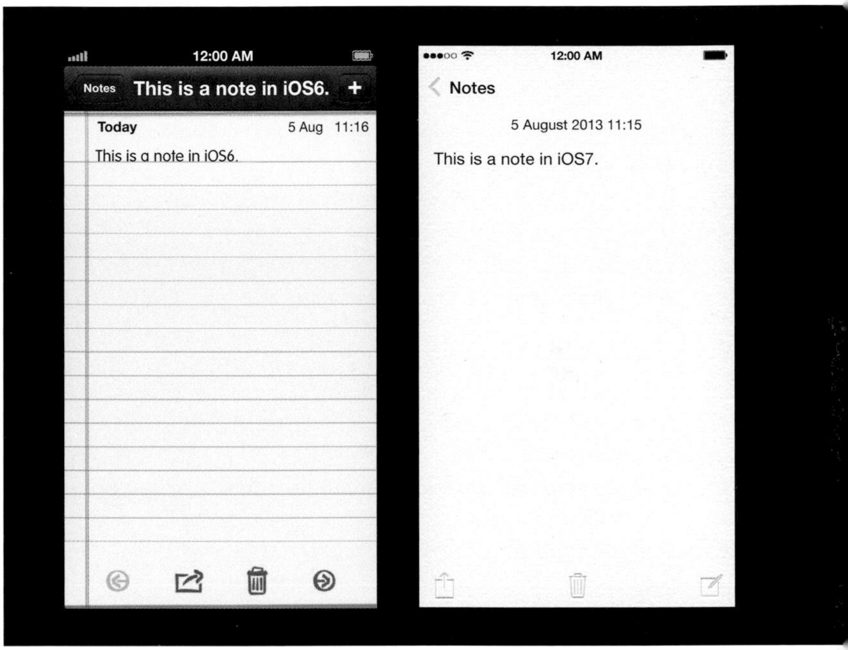

At Apposing, we favor the native app for the majority of our applications. With constant updates to operating systems and mobile functionality, it's the native application that remains at the forefront of mobile advancement. Without immediate access to unique features, hybrid applications are always playing catch-up to their native counterparts and currently offer a much slower experience for the user. It's the creation of the native application that we'll be exploring in this book.

Characteristics of a Successful App Designer

SUCCESSFUL APP DESIGNERS SHARE A SET OF COMMON TRAITS:

They keep it simple: When it comes to app design, it's best to let your app's primary function shine through. If you can get away with removing a stage of a process, do it. Simplicity is key, and your app's users will thank you for it.

They're thoughtful: By putting yourself in your app user's shoes, not only will you get a better understanding of the process and pitfalls associated with it, you might find an opportunity to make your app stand out. The British Gas application solved a problem many users weren't aware they had. The user journey took into consideration the way that people would use the application. With the majority of gas and electric meters in the UK housed in dark cupboards or basements, the simple addition of a torch to the meter reading screen made a monthly task significantly easier, and ensured that users returned to the app, continually exposing themselves to the app's other functions and learning more about the brand's services.

They're unique: There are around a million apps in the App Store, with more than a billion having been released since the App Store opened. According to analyst house Canalys, just 25 developers took home half of all app revenue from the two dominant U.S. app stores, Apple's App Store and Google Play, over a 20-day period in November 2012. If you want to stand a chance of your app being discovered, never mind successful, you'll have to battle against similar apps with almost identical keywords (something we'll explore in chapter nine). The only way you can ensure that yours stands out (huge marketing budget aside!) is to make it the only one of its kind. Or, if you're improving on an existing app, make sure you add your own stamp. Find the factor that sets your app apart from the rest and build on it.

They're approachable: If you're going to be using an app daily, a condescending tone, boring voice, or sickly color scheme can lead you to press delete before the app has been integrated into your life. A fun, friendly, and efficient app will win over the hearts of users quicker than one that tries to be superior.

They're interested and interesting: If you don't take the time to fully understand how people could use your app then you're missing a trick. Great product design stems from extensive research. The developers watch how people interact with products and their environment, noting the order in which they do things and the problems they come up against. Good app design should do the same. If you can't be interested in your chosen field then your app will almost certainly not be useful or interesting.

They embrace the future: The future is your friend, and you can help to shape it! Advancements happen daily; the best apps take advantage of these developments and integrate them into their design. The pace at which technology moves forward is exciting. A good app designer will embrace these changes, not fear them. If you don't look to the future with your designs, your apps will date in no time at all.

Good app designers have a knack for addressing problems that the public never knew that they had, enhancing lives in the process.

How to Transition from Web or Print Design to Mobile Design

If you're serious about moving into app design, then now is the time. As the app market proliferates, the opportunities for designers are increasingly daily.

For those already involved with web design, the transition will be easier as certain principles are shared between the two disciplines. Debate still continues in the industry on whether both disciplines actually fall into the category of "interaction design" (IxD), but for the purposes of this book, we'll keep them separate.

The advent of the hybrid app also offers web designers an easier "in" to the world of app design than their native counterparts, as programming languages such as HTML, CSS, and JavaScript can all be used to create beautiful applications that can then be sold through the App Store. Hybrid apps don't require web designers to have in-depth knowledge of Objective-C. Native apps, the solution which currently offers the richest and fastest user experience, do require knowledge of Objective-C, so if you're coming at app design with no plans to learn the language, you'll either need to get some training in Objective-C, use a template, or have a developer or programmer on hand to do the bits you can't.

Launching into a career in app design from a print background is a little harder as you may not be as well versed in the conventions and psychology that apply to both app and web design. However, this book will take you through the most important of them.

Ultimately, everyone needs great design, and if you're able to incorporate your eye for design with a sharp understanding of your target market, then you're on your way to designing a great app.

If you're able to incorporate your eye for design with a sharp understanding of your target market, then you're on your way to designing a great app.

TOOLS YOU'LL NEED

Thankfully, many of the tools you'll need in app design are the same as for web. Though this book won't teach you how to use the programs listed here, it will teach you how to integrate the assets you've created into a usable application.

A Mac: Though it seems pretty self-explanatory, many people still think they can run Xcode on a PC. There are workarounds out there, it is by no means impossible, but the alternative routes won't be covered in this book. Oh, and you'll need to be running the latest OS X (operating system).

Apple Developer License: You need a developer license to distribute (i.e. release) in the App Store and to test on devices. You can test on the simulator without a license.

Xcode: This piece of downloadable software can be accessed for free through the Mac App Store. This is where you'll spend time writing code in Objective-C.

Adobe Photoshop or equivalent: You're a designer, so you probably already have access to some sort of image creation and editing software.

Adobe Illustrator or Corel Draw: iPhones come in all shapes and sizes. Some designers wish to create their assets as vectors to account for sizing up and down without their components pixelating.

Prototyping and planning tools: These don't have to be programs, though there are some great ones out there. We use everything from paper templates to mocking straight up in Xcode. We'll cover these in depth in chapter three.

CHAPTER TWO

HOW ARE APPS MADE?

Though the idea for your first app may have hit you outside of a planned brainstorming session, taking that idea to the next level should take a more structured and methodical path. This chapter outlines how iPhone apps are made. As you'll come to learn, an app idea is not a plan; the idea is the merely the starting point. From defining exactly what it is that your app does to testing your idea within an inch of its life, we've mapped out our tried and tested methods of creating a fully functional app, leaving nothing to chance. You'll also need to make choices such as whether to design from scratch or to enlist the help of a template; this chapter gives some food for thought on which is the best for you.

App Design and Development Process

While there is no right or wrong way to design an app, you'll find that each development studio or designer will favor a particular approach. There are numerous well-documented methods of developing software. It's important that you find the process that works for you and gets the best results for your project.

Our Method

Here at Apposing, we've taken bits of many processes to create the method that's right for us. We'll go more in depth with each stage in chapters three and four, but until then, our usual journey goes a little like this.

Sticking to a development plan will stop the process from overwhelming you, allowing you to focus on design.

VISION

Quite simply, during this stage we define exactly what it is that an app does and who it's for.

We start every project with clear goals and establish a primary function. Defining exactly what it is that an app does and keeping that thought with you throughout the process is something that can prevent a lot of unnecessary work further down the line. The very act of defining and writing down your idea will keep you on the right track. Ask yourself what functions it will use and to what purpose. Though later stages may well add to or evolve your idea, keeping what you want to achieve constantly in mind will stand you in good stead.

As well as knowing what your app does, it's important to know who it's for. Who is it that you envisage using your app and in what circumstance do you imagine them using it? While everyone thinks that their app idea is one that will revolutionize a task or take the App Store by storm, reality paints a very different picture. Defining your audience is key for many reasons. Functionality, market research, and app marketing will all take a very different turn depending on who it is that will be using your app.

RESEARCH

Always do your homework. Though after defining your idea, you might want to rush headfirst into developing it, there's a rather important stage to come first. You need to know what's out there. Missing out this stage is a bit like starting on a recipe without knowing if you have all the ingredients.

Smart designers will research their idea and potential market extensively. With thousands of apps entering the App Store monthly and a finite number of smartphone functions, there's bound to be some crossover with other applications. Unless you've invented something entirely revolutionary, you will definitely have competitors.

PLAN

This stage is imperative to a clean and smooth user experience (UX). If you've done your research properly, you should now know an awful lot about both the market and your target audience; the how, when, why, and where of using your app. Using the phone functions available to you (there's a full list in chapter five), the designer needs to identify exactly what it is that their app needs to do and the phone functions it will need to achieve this.

It's only then that you can effectively plan your user journey. Using the app's primary objective, the designer will visually prototype the app. Whether you use paper, a template, or Photoshop to realize your idea, just getting it down and laid out in front of you may reveal features you've missed, problems you may face in development, and things you could leave out. From creating the sitemap to wire-framing your product, we cover these stages in depth in chapter six.

REGISTER WITH APPLE

If you haven't done so already, you'll need to register yourself as an Apple Developer before you can submit to the App Store. It costs $99 for a one-year license as a registered developer. The developer portal contains a raft of analytics and other tools to help you on your way.

There's also a set of guidelines that you'll need to adhere to in order to make sure your app is approved by Apple. By getting to grips with these now, you're preventing heartache further down the line. You can find guidelines for all the areas you'll cover here: https://developer.apple.com

DESIGN

With your wireframes and user flow locked down, the next stage in the process is the realization. Whether you're designing from scratch or using a template, your prototype is brought to life in Photoshop or an equivalent program at this stage, adding images and textures as your app's UI takes shape.

If you suddenly realize that your app is lacking a vital function, or decide that your menu isn't easy enough to navigate, then it is during this part of the process that you change your app's design for the final time if you want to stick to conventional methods of development.

MARKETING MATERIALS

Though the app itself may be finished, there's still a number of things you need to consider, from the App Store icon and screenshots to the text which accompanies your app in the App Store. Before you can submit, you'll need to have these things ready. You'll find out more about this in chapter nine.

SUBMIT

When your account is active, you'll upload your app files, screenshots, icons, and marketing materials before submitting to Apple for approval. This stage can take anywhere from a few days to a few weeks. If approved, the app doesn't go live until you publish it. You can set it to publish upon approval, but if you've got a marketing plan in place you probably don't want to choose this option. You can set a release date in the future for your app to go live if you wish.

BUILD

If you're not coding your app yourself, then this is where you hand your baby over to a programmer. If you are, then your Objective-C journey begins here. Changes made to the functionality design at this stage are not irreversible or completely out of the question, but will have a massive impact on your app's build time and should be avoided if at all possible.

TESTING, TESTING

Quite self-explanatory, it's important that you test your app on the various models of iPhone and iPod (and iPad if you're going down that route). By allowing people who are unfamiliar with your product to get their hands on it, most problems are likely to surface now. We use test scripts to make sure that our apps are free from errors. A test script is a set of instructions which the tester follows to make sure that all areas have been tested and that all navigational methods are fully functional. Any glitches or design errors can be fixed at this stage, and then tested again before the final files are created.

MANAGE

If you've reached this stage, your app is now live in the App Store. This isn't where the story ends. There may be bugs, there will be fixes, and, ideally, you've planned a future for your app that contains updates with new content or extra features. Like the app itself, these changes are submitted for approval before going live.

So there you have the basic stages of app development. We'll look more closely at each of the stages in the next few chapters.

The Different Types of Apps

If you've decided to make the move into iPhone app development, it's very likely that you have an iPhone. At this very moment, a multitude of apps are probably living on your home screen, the nature of which leads people to group their apps together into handy little categories. These make apps easier to find and organize. Utilities, games, arcade, photography, lifestyle—the list of categories goes on and on. Types of applications, however, can be broken down into a much smaller list.

***Fundamentally, you can break apps down into two categories:* FUN *and* FUNCTIONAL.**

On a very basic level, fun apps are game and entertainment applications. Functional apps are tools and utilities; apps such as Mail, or apps that help you to complete a task or collate information.

Though that seems a little simplistic, from a design point of view it makes sense. The way we approach UI design for each type will differ slightly.

Fun apps, with particular reference to games, aim to capture the user's interest for a longer time than their functional counterparts. Attention-grabbing and graphically richer than their functional brothers, fun apps can take more chances with their design. Fun apps rarely have complex navigational systems, often taking the user straight to the focus of the fun. Perfect examples of fun apps are titles such as Temple Run or Letterpress.

While every app's UX needs to be free from clutter, unnecessary extras, and screens, functional apps are there to fulfill a purpose. Apple's Mail app is the archetypal example of a functional app. While it's not the prettiest thing we've ever come across, it does its job and it does it well. Apps like Mail feature content that pushes a lot of information onto the user. By keeping the design and functionality basic, the user isn't overwhelmed and is able to sort through the data easily. Adding logos, images, and other extras, though possibly adding visual appeal, would dilute the app's simplicity and would make it harder to use. Placing purpose over style, functional apps are just as their name suggests.

But can an app not be fun and functional?

Fun and functional often overlap, as our diagram shows. Placing your app idea onto the chart will give you some indication toward the direction your design should take. From using Apple's default graphics to putting function over design, these elements are all things to consider when planning your app.

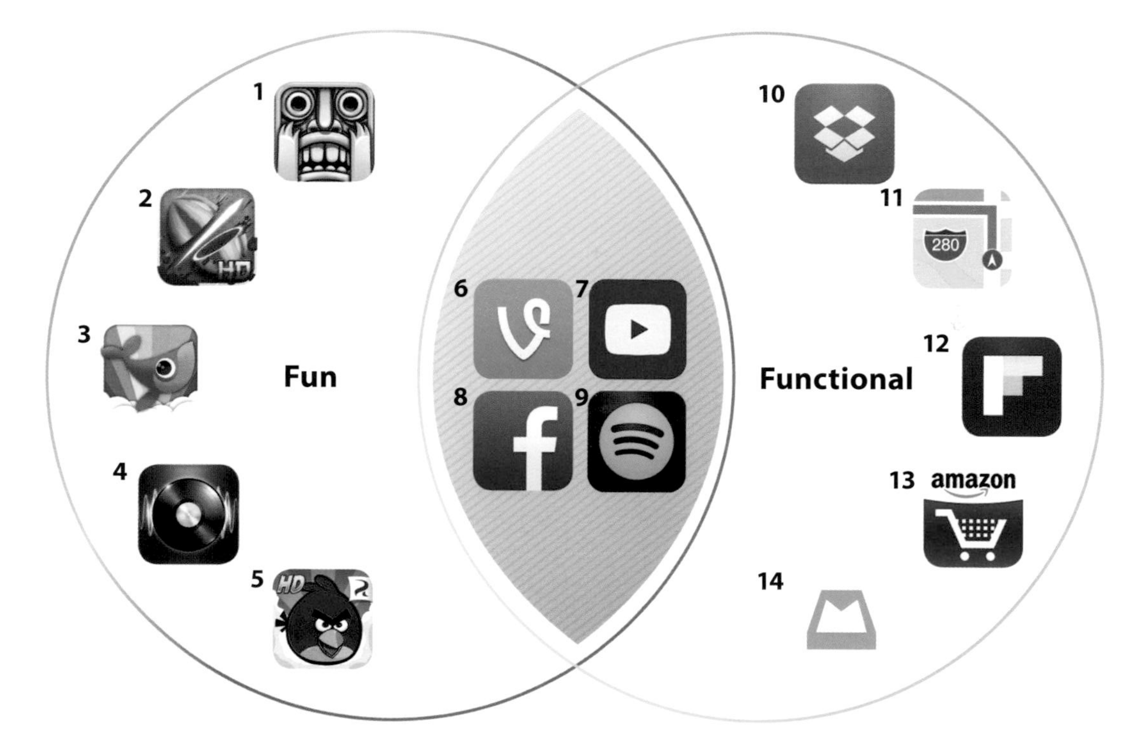

1	Temple Run	6	Vine	11	Apple Maps
2	Fruit Ninja	7	YouTube	12	Flipboard
3	Whale Trail	8	Facebook	13	Amazon
4	DJay	9	Spotify	14	Mailbox
5	Angry Birds	10	Dropbox		

Programming Languages

As mentioned in chapter one, native iPhone apps are created using a programming language called Objective-C. The language is used through the Apple program Xcode. When you register for an Apple developer account, you'll get access to the software developer kit (SDK) for free. The software is downloaded and then installed on your Mac, which will need to have been updated to the most current OS X operating system.

The rise of the hybrid application, however, means that apps with almost all native functionality can now be created using HTML, CSS, and Javascript via a third-party framework. Native functionality refers to functions including the phone's camera, GPS, and accelerometer, and native software which allows you to access features such as the phone's photo and video libraries, contacts, and calendar.

Tools such as Phone Gap allow the user to build an app using traditional methods and languages. The finished product is then "wrapped" in a native shell by the program to allow it to be submitted to the App Store.

Benefits of this method mean that your app can be reversioned for Android and Blackberry platforms with minimum rewriting of the code. The downsides of using this method include slower performance, and with each update to OS X, the third-party frameworks will need to develop a new workaround, so access to newer functions is delayed until the software catches up. As with all software, there will be extra bugs to contend with which can add extra time to the testing process. Different frameworks will have their own issues, with workarounds sometimes documented, sometimes not.

It's also worth noting that even when using a third-party framework, some knowledge of Objective-C will be necessary to submit the app for approval.

As mentioned, Apposing develop natively, meaning we write our iOS apps using the coding language specific to the platform, Objective-C, and our Android apps using Java. We do this using the software developer's kits (SDKs) provided by Apple and Google.

Developing applications natively allows Apposing to adopt the latest developments as soon as we can, ultimately passing on the benefit to our customers.

Objective-C is a universal language; programmers speaking any language will use the same set of terms.

GEITF.xcodeproj — ProgrammeViewController.m

GEITF: Ready | Today at 16:36

GEITF › GEITF › Programme › ProgrammeViewController.m › -tableView:cellForRowAtIndexPath:

```
    legendViewController.modalTransitionStyle = UIModalTransitionStyleFlipHorizontal;
    [self.navigationController presentViewController:legendViewController animated:YES completion:nil];
}

#pragma mark - Table methods

- (NSInteger)numberOfSectionsInTableView:(UITableView *)tableView {
    return 1;
}

- (NSInteger)tableView:(UITableView *)tableView numberOfRowsInSection:(NSInteger)section {
    return [chosenDayEventsArray count];
}

- (UITableViewCell *)tableView:(UITableView *)tableView cellForRowAtIndexPath:(NSIndexPath *)indexPath {

    static NSString *CellIdentifier = @"Cell";

    EventCellView *cell = [tableView dequeueReusableCellWithIdentifier:CellIdentifier];

    cell.accessoryType = UITableViewCellAccessoryNone;

    if (cell == nil) {

        cell = [[EventCellView alloc] initWithStyle:UITableViewCellStyleDefault reuseIdentifier:CellIdentifier];

        UIView *v = [[UIView alloc] init];
        v.backgroundColor = [UIColor colorWithRed:255/255.0f green:254/255.0f blue:187/255.0f alpha:1];
        cell.selectedBackgroundView = v;
    }

    //Event *aEvent = [eventsArray objectAtIndex:indexPath.row];
    Event *aEvent = [chosenDayEventsArray objectAtIndex:indexPath.row];

    cell.colourView.backgroundColor = [self colourForVenue:aEvent.venue];
    cell.eventNameLabel.text = aEvent.title;
    cell.eventTimelabel.text = [self formattedEventTime:aEvent.startTime endtime:aEvent.endTime];

    return cell;

}

-(void)tableView:(UITableView *)tableView didSelectRowAtIndexPath:(NSIndexPath *)indexPath {

    [tableView deselectRowAtIndexPath:indexPath animated:YES];

    Event *chosenEvent = [chosenDayEventsArray objectAtIndex:indexPath.row];

    EventDetailViewController *eventDetailViewController = [[EventDetailViewController alloc] initWithNibName:@"EventDetailViewController" bundle:nil];
    eventDetailViewController.event = chosenEvent;
    eventDetailViewController.navigationController.navigationItem.title = chosenEvent.title;
    [self.navigationController pushViewController:eventDetailViewController animated:YES];
}

-(NSString*)formattedEventTime:(NSString*)startTime endtime:(NSString*)endTime{

    NSString *start = [startTime substringToIndex:[startTime length] - 3];
    NSString *end = [endTime substringToIndex:[endTime length] - 3];

    return [NSString stringWithFormat:@"%@ - %@", start, end];
}

-(UIColor*)colourForVenue:(NSString*)venue{
```

Custom-designed Vs. Template Designs

If you have no previous experience in designing apps, you may want to start out using drag-and-drop UI templates. These may have the navigation and transitions already set. While at Apposing we favor designing custom graphical user interface (GUI), there are times where using a template is more time or cost effective, with little impact on the final product.

Templates are easily customized, though the extent of the customization is limited and varies on a case-by-case basis. It's worth bearing in mind that Apple's UI elements (see page 35) are also easily customizable. So if you decide that you fancy going template-less, you can still use custom elements and tailor factors such as color and tab icon images (there are still limitations) to make the standard system controls match the branding of your app.

Those keen to muddy their hands with Objective-C are also likely to steer clear of using templates.

If at the end of the book you're still unsure of navigational methods, or if time is of the essence, then a template may be for you. Alternatively, there are a bunch of much emptier templates for those wishing to design their own user interface but are just starting out. These come perfectly proportioned for the various screen sizes, and some have rulers for the status bar and other Apple elements. Designers can use these templates for practice and to get to grips with the dimensions required for the various iPhone models.

Whether you're using a template or otherwise, it's essential to familiarize yourself with the Apple Interface Guidelines. While it certainly won't be the most interesting read of your life, fully understanding what you're up against will help you navigate your way through the project. Also, you'll be provided with the exact specifications required and it comes complete with some very sound advice.

It's worth pointing out that using a template doesn't mean you won't have to use Objective-C at some point. Most templates offer you a design and design only. You'll still need to code the design, or alternatively get a developer to do it for you.

UI DESIGN TEMPLATE:

1. **Your company or product doesn't have strong branding:** If your branding isn't all that relevant to the product, using a template won't detract from what people expect from you. If you've spent years cultivating an image, logo, color scheme, and a custom font, and more time designing a website that brings the whole thing together, using a template which doesn't quite gel would be a waste. For those with weaker branding who wish to go it alone, custom GUI could be an opportunity to develop a strong look for your product.

2. **Limited timescale:** Sometimes there just aren't enough hours in the day. Using a template removes a whole chunk of the design process, and ideally offers you a template fully compliant with the Apple guidelines, lowering the chances of your app being rejected upon submission.

3. **Smaller budget:** Time is money. By cutting out much of the design process, you'll save yourself a pretty penny.

To Pay or Not to Pay?

If you've settled on using a template, then the raft of them available online may leave you scratching your head. Like WordPress and Tumblr themes, app templates come in free or premium options. You may be wondering what the differences between the two are and the benefits of parting with your cash for the premium templates.

Unfortunately, there are simply too many templates available to be able to offer advice on each one and say if the price tag attached to it is fair, but here are some things to consider which should help you decide.

Research the vendor and make sure you check any feedback before you part with your hard-earned dollars. It's also worth checking out the template rules for commercial use before you purchase—if you're using the same template multiple times, it's worth noting that some template licenses are restricted for use in a maximum of two apps live in the App Store. Also, make sure you check exactly what you can customize and what you can't before setting your heart on a template.

While there are certainly cases in which using a template would be the best course of action, there are an infinite number of reasons to use a custom-designed GUI. To achieve any level of success, you'll need to stand out from the ever-increasing number of apps available, and a beautifully executed and unique custom design is one way to differentiate yourself. if you have a keen eye for design and you want to create something which showcases your artistic flair and demonstrates a truly unique product, then a template is probably not for you.

A beautifully executed and unique custom design is one way to differentiate yourself from the thousands of apps in the App Store.

REASONS TO USE CUSTOM DESIGN:

1 **Strong design skills:** If you're a particularly awesome designer with a superb understanding of your audience, as long as you have a way to code your design, we'd advise you to get stuck in.

2 **Larger budget:** We won't lie, custom design costs more, both in time and money. Man-hours are expensive, but if you've got the budget to cover it and it's something unique to you that you're after, then custom is what you need.

3 **Unique needs/user journey**: If your app is particularly novel or the UI requires something a little out of the ordinary, it's unlikely that you'll find it on the shelf. Custom design offers you the opportunity to get exactly what you need with no compromise.

4 **Longer project timescale:** If time isn't your enemy, then it's worth considering custom over a template. Though it may take longer, the end result will be truly unique.

When to Use Apple UI Elements

Standard Apple UI elements are app components such as menus, buttons, status bars, navigation, table view, and icons.

Initially, when app development was in its infancy, Apple pushed developers toward using standard Apple elements so that app users would benefit from a uniform experience and to quicken the speed of app development. Standard Apple UI elements can be found in the iOS Developer console, and feature everything from navigation bars to switches and sliders. The elements, when used correctly, make it easier for app developers to conform to the Apple guidelines and build apps quickly. That said, app development has moved on a lot from then, and many people's designs are made all the better by their custom elements.

EXAMPLES OF CUSTOM DESIGN:

- Camera + looks like the standard camera app, but its custom UI offers many more features. A small settings menu next to the central button reveals handy functions such as the timer, stabilizer, and the option to shoot in bursts. All functions can be accessed on the action screen, which makes for a faster and smoother experience, handy when the perfect photo opportunity is lost in an instant.

- Tweetbot's tab bar has much more functionality through its custom UI than Twitter.

- Vine offers users a much easier way to record than the standard video recorder through its "touch and record" UI design.

It's important to ask yourself whether your design will truly benefit from custom-designed navigation, which will undoubtedly add extra time and expense to the project.

12:00 AM
Search
apposing
APPOSING is not following you.
APPOSING
@apposing
Liverpool & London
http://www.apposing.co.uk
We develop ideas-led mobile apps & forward thinking technology for Nando's, Carphone Warehouse, CSL Sofas, Unilever, BBC & Boots. An Accelerate250 company.
Last tweet 27 minutes ago
Tweets 13,886
Followers 2,707

Tweetbot

Apple video camera

12:00
NEW BOOKING
From: Apposing Ltd, 46 Jamaica...
Postcode: L1 0AF
To: FACT, 88 Wood Street, Liver...
Postcode: L1 4DQ
Pick-up Time: 19.00
Pick-up Date: 20 Aug 2013
Passenger Name: Andy
No. of Passengers: 2
Vehicle Type: Saloon Car

Cabfind IOS 6

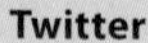

Twitter

Vine

12:00 AM
Back
NEW BOOKING
From: Apposing Ltd, 46 Jamaica St...
Postcode: L1 0AF
To: FACT, 88 Wood Street
Postcode: L1 4DQ
Pick-up Time: 19.00
Pick-up Date: 20 Aug 2013
Passenger Name: Andy
No. of Passengers: 2
Vehicle Type: Saloon Car

Cabfind IOS 7

As with template versus custom design, the decision to use the standard Apple elements should be considered on a case-by-case basis. It's important to ask yourself whether your design will truly benefit from custom-designed navigation, which will undoubtedly add extra time and expense to the project.

Apple's standard elements have progressed since their inception. APIs (application programming interface) for iOS 5 and higher allow customization of most UI elements.

The standard elements debate also crosses over into app maintenance and in future-proofing your app. Standard UI elements will automatically receive appearance updates, while completely custom UI elements will not. Customized standard elements do benefit from the automatic updates, so score a few more plus points.

Remember, standard elements for iOS 6 and iOS 7 differ in appearance quite dramatically. If you're planning on supporting iOS 6, then you won't be able to use all of the iOS 7 elements in your design.

THINGS TO CONSIDER WHEN DECIDING ON CUSTOM OR STANDARD UI ELEMENTS:

1 **Budget:** Choosing custom elements will add cost to the build. If you have a small budget and the custom UI benefits nothing but your vanity, it might be better to use Apple's standard elements and customize them.

2 **Branding is unnecessary:** If you're making a functional app which has all the emphasis on the data it features, as previously discussed, branding might be a hindrance to the UX. The same goes for custom elements. If adding custom elements will disrupt the view for the user, stick to standard.

3 **Phased roll-out:** If you're planning a phased roll-out with completely custom UI elements, if there's an OS X update mid-way through, you'll need to modify your custom elements accordingly. Using standard or customized elements will avoid this step through automatic updates.

CHAPTER THREE

DEFINING THE APP'S PRIMARY TASK

We've touched on this before, but some of the most important work when designing your application happens before you even put pen to paper. Market research (which we get to in chapter four) will never be successful if you're not looking for the right statistics, trends, and facts.

The only way to guarantee that the rest of the process will run smoothly and with minimal hitches is to define in great detail just what it is that you're making.

From an easy-to-use and well-spaced user interface (UI) to an intuitive and enjoyable user experience (UX), by getting this step right you can ensure you're hitting the right bases during the rest of your app's development.

What Task Does Your App Accomplish?

We are approached by hundreds of entrepreneurs, business owners, and respected industry figures on a monthly basis, each believing that their app idea is the golden goose. However, when we ask them to get to the nitty gritty of what their app does, the concept often loses much of its shine.

We listen to every idea that comes our way, and advise each client on an individual basis. If we don't believe that an idea or vision will fly, we say so, whether the idea would be worth a lot of money to us or not. As stated in the previous chapter, to develop a killer app is to develop something very personal to the user. If an app idea doesn't cut it at this stage, it should result in a trip right back to the drawing board, no arguments.

The same goes for designing your own app. Though defining its purpose sounds relatively simple, there are many factors that need to be considered at this early stage. Some functions need to be built into the foundations of your app and may change the final layout and design, so it's vital that you finalize these details now.

If an app idea doesn't cut it at this stage, it should result in a trip right back to the drawing board, no arguments.

The likelihood is that you have a few app ideas churning away in the back of your mind already. Either you've found a task that there wasn't already an app for, or you've found an existing one that's so lacking in functionality, you've been inspired to take the idea back to the drawing board and fix it. Whatever stage you're at, write down every idea that you have. There's nothing worse than dreaming up some great concept only to forget it at a later date.

If you ever encounter a problem, or you find yourself thinking "I wish there was an app for that," the chances are that others are feeling the same. Make sure you note it down. Check to see if there is an app which fulfills this function already out there, and if there's not, you should consider making it.

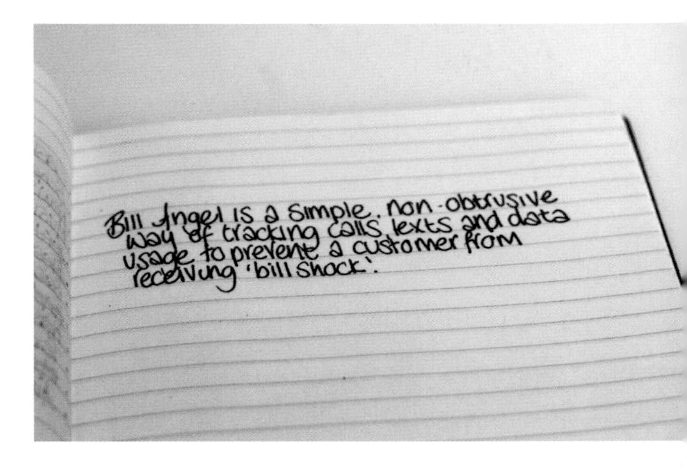

Your Mission Statement

When you've got an idea that you're excited about, it's time to take it to the next level. No matter what extra functions are added during the planning process, your app will have a primary function, the crux of your idea. The design process truly begins with you writing down your mission statement. In 25 words or less, write down exactly what your app does, what makes it different, and who it's for. For example, our statement for our app for CSL, a UK-based couch specialist, would go a little something like this: "In Your Room allows the user to see CSL's entire catalog of couches in their own living room through their camera using augmented reality."

The finished mission statement should be at the forefront of the majority of your design decisions. This will help ensure you don't veer off in an alternative direction by adding unnecessary functions, or allow your UI to obstruct the user from getting to the task at hand.

Getting Feedback

With the mission statement ready, it's time to get feedback from others. Starting with family and friends, you need to gauge reaction to your idea. Ask for complete honesty and be prepared to take rejection well. Having people whose opinion you can rely on is priceless and could end up saving you a lot of time and money in the long run.

If at this point your idea still seems viable, you might want to run it by an expert. Though sharing your idea might seem like crazy talk, it's useful to know that the industry values the execution much higher than the idea; a 20/80 split in fact, that's 80 percent for the development. Talk to an app developer or someone in a similar position. They'll be quick to tell you whether it's worth pursuing, and at the very least will explain why they feel the idea will or won't be a success.

Much like an elevator pitch, you should be able to explain your app idea to a stranger in 30 seconds, winning them over in the process. If you can't manage this, your idea is almost certainly too complex.

If you don't have access to an expert, get tough on your own idea and put yourself in your potential user's shoes. Be realistic and brutally frank.

THINK ABOUT:

1. How often will you use it?
2. How much will it improve an everyday task or problem?
3. How engaging will the experience be?

If, even when being your harshest critic, you still think you're on to a winner, then it's time for some other concerns.

More Things to Think About

Deciding whether an idea is worth pursuing isn't just a matter of functionality. There are other factors to consider, such as whether you want or need your app to make a profit, and how much time you're willing to spend on it.

ADDITIONAL THINGS TO CONSIDER WHILE YOU DEFINE YOUR APP:

1. What do you expect or need to achieve? What are your goals? While attempting to reach the top of the app charts with your first app is impractical, setting yourself achievable goals will give you more of an indication of how much work will be involved before, both during and after your app is finished.

   If your app is just for fun then your goals will be looser, but if your app is to act as another arm of your business's marketing efforts then you must set yourself goals and targets to be able to effectively measure your app's success.

   If you need your app to make money, then you're going to need to factor a monetization plan into the development process, either through upgrades, advertising, or other means.

2. How much of the work will you need to outsource? Can you code? Will you be able to design a UI to a high enough standard? Can you create any animations or other assets that are required? If the answer is no to any of those, then you're going to have to outsource. Finding the right person will take time and it will cost a lot, especially on the development side.

3. How much time can you spare? How much time can you commit to this project? This applies to the build and beyond, and could affect the app's primary task. If you don't have enough time to commit to the app's upkeep, answering users' questions, responding to requests, and developing new versions, then factor this into the app's design. If your time is scarce, then the app's functionality may have to reflect this, as a more complex app generally equates to a longer build time.

4. What's your budget? Your budget needs to cover all of the factors you've answered in the questions above. Though your idea may be a winner, if you're lacking the skills required, you're going to need money or someone who believes in your idea as much as you do.

Removing Function Overload

With your mission statement as your starting point, you'll need to identify the important tasks and factors associated with your app's central theme and decide which you'll need to include.

For example, during the development process for Clashfinder, an app for festival-goers across the UK, to feature many of the biggest events over the summer period, we loosely decided that the app should allow users to enter the bands that they were going to see, with the app finding any clashes or overlapping acts in their planned activity. After researching other festival apps, we found shared and common functions in many of the current offerings which we added to a list of possible functions. Initially, the list of potential functions looked like this:

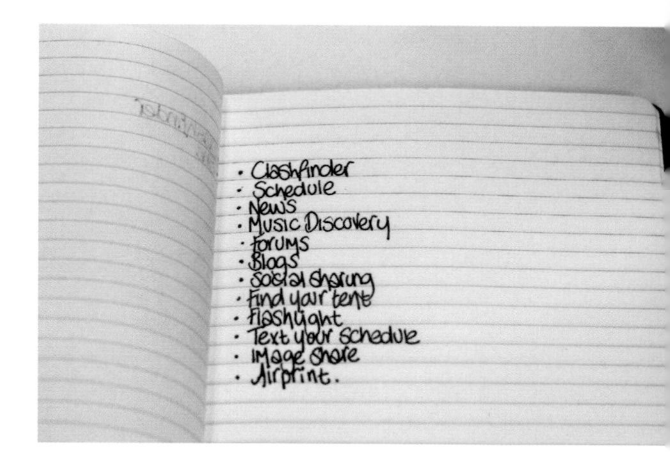

- Clashfinder: The app's primary purpose.
- Schedule: Goes hand-in-hand with the Clashfinder.
- News: Updates, announcements, and more.
- Music Discovery: Images and videos of the acts. For users unsure of an act's output, Music Discovery allowed them to explore the act's sounds via videos and images.
- Forums: Giving festival-goers the opportunity to chat about the event, arrange meet-ups, and share information.
- Blogs: Features on the upcoming event.
- Social sharing.
- Find Your Tent.
- Flashlight.
- Text your schedule and clashes to yourself or a friend.
- Share your plans and images on social networks.
- Airprint: Print out your schedule and clashes before you leave.

While this list of potential functions seems like the whole package, in app form, it would be a little overcrowded and confusing for the user. Though pushing boundaries is normally considered a good thing, when designing your app it's better to make the most out of a few functions rather than to stuff in everything possible. The saying "jack of all trades and master of none" is very fitting in the world of app development. We've been conditioned to think that more is always better, but in terms of developing the scope of an app, the UX will almost always be hindered by function overload. Your app needs to be the master; it doesn't need to be everything to everyone.

While all the functions in Clashfinder listed might be useful or fun, they're not all necessary.

As well as focusing on the app's primary purpose, the aim for Clashfinder was to produce an app with easy-to-use, high-quality functions, rather than to overload it with several poor-quality features. Feature overload will often lead to a lesser experience for the user, with rushed

What was removed:

- Forums
- Blogs
- Social sharing
- Find Your Tent
- Flashlight

What was kept in:

- Clashfinder
- Schedule
- News
- Music Discovery
- Text your schedule and clashes to yourself or a friend
- Airprint

development time being one of the side effects. When there are apps such as Torch already available in the App Store (and it's available in the control center in iOS 7 without having to download anything), why waste space in your own app for a feature which is unlikely to perform the function better? The decision to remove many of the features was based upon this assumption. The features that were kept in were ones that added to the primary function; the ability to text or print clashes supplemented the primary function, and Music Discovery aided people's schedule decisions, as did the news. The features that were taken out of the list, though useful in their own right, could have diluted the app's streamlined experience by congesting the user flow, without actually adding much to the app's primary purpose. While the myriad of features you have planned for your app may seem exciting to you, bombarding your users with choices in such a tiny piece of real estate can be confusing and overwhelming. To be able to use an app intuitively, without any real instructions, the user journey needs to be clear and simple.

Refining Your App's Features

If you have got to this stage, you are likely to have a central feature, or a few features in mind for your app. Instead of asking yourself what else you could add, ask how you can make the absolute most out of the features that you have settled on.

ASK YOURSELF:

1. How can I make sure the main function is easy to use?
2. Will people understand instinctively how to use it?
3. How quickly will the user be able to achieve their objective?
4. How can we make it clear to potential users that this is what this app does and that it does it better than any other?

If your app is laden with vaguely useful functions, superfluous to the central purpose, the chances of it being easy and quick to use and becoming an everyday tool in your user's arsenal are slim.

By now you should have a clear idea of what your app is and where it's going. You should also know what your expectations are for its success and the bare minimum that you need it to achieve.

CHAPTER FOUR

MARKET RESEARCH

By looking at many of the failed apps in the App Store, it's obvious that many designers decided to forego the market research. Instead of getting to know the market, they launched headfirst into an app design which demonstrates a complete lack of understanding of its target audience. This stage in the process will help to nurture your idea into a viable app concept by identifying gaps in the market, and strengths and weaknesses of existing products. This chapter will outline the importance of getting to know your audience, and also demonstrate how a similar app doesn't have to mean the end for your idea.

Concept Research

Good app design isn't a "one size fits all" process. This stage of proceedings allows you to get inside the head of your target audience, going on to craft the app they actually want, rather than the one you think they need.

If your idea for an app stems from someone else's awful effort, you've already latched on to the notion that finding an app with the same premise as yours needn't mean curtains for your concept.

While we all want to be the trailblazing future-thinker, many brands, developers, and designers have made their mark by improving on an existing product. Being able to identify exactly why an app has failed and being able to further that recognition with solutions that make the app both useful and useable is a veritable skill in its own right.

That being said, the fast pace of iOS updates, developments, and new phone functions means that there is room for absolute innovation too. The important thing to recognize when it comes to app market research is that, in terms of design, there will always be some overlap with other applications already out there. Much like a musician knows that there are only a finite amount of notes, there is a limited feature set available on an iPhone.

Though you may feel that your idea is one of a kind, that's no excuse to skip this step as there's still much to learn from apps that share the same features, or break boundaries in terms of custom UI or UX. This chapter will outline how to go about strengthening your idea by learning from those who've gone before. It doesn't make your idea weaker; it is application evolution.

We'll also look deeper at who you're designing for and how it will, and should, affect your final design. App design, like web design, is no longer just a matter of esthetics; UX should also be a central focus. Understanding your market is everything—get it right and your audience will thank you for it.

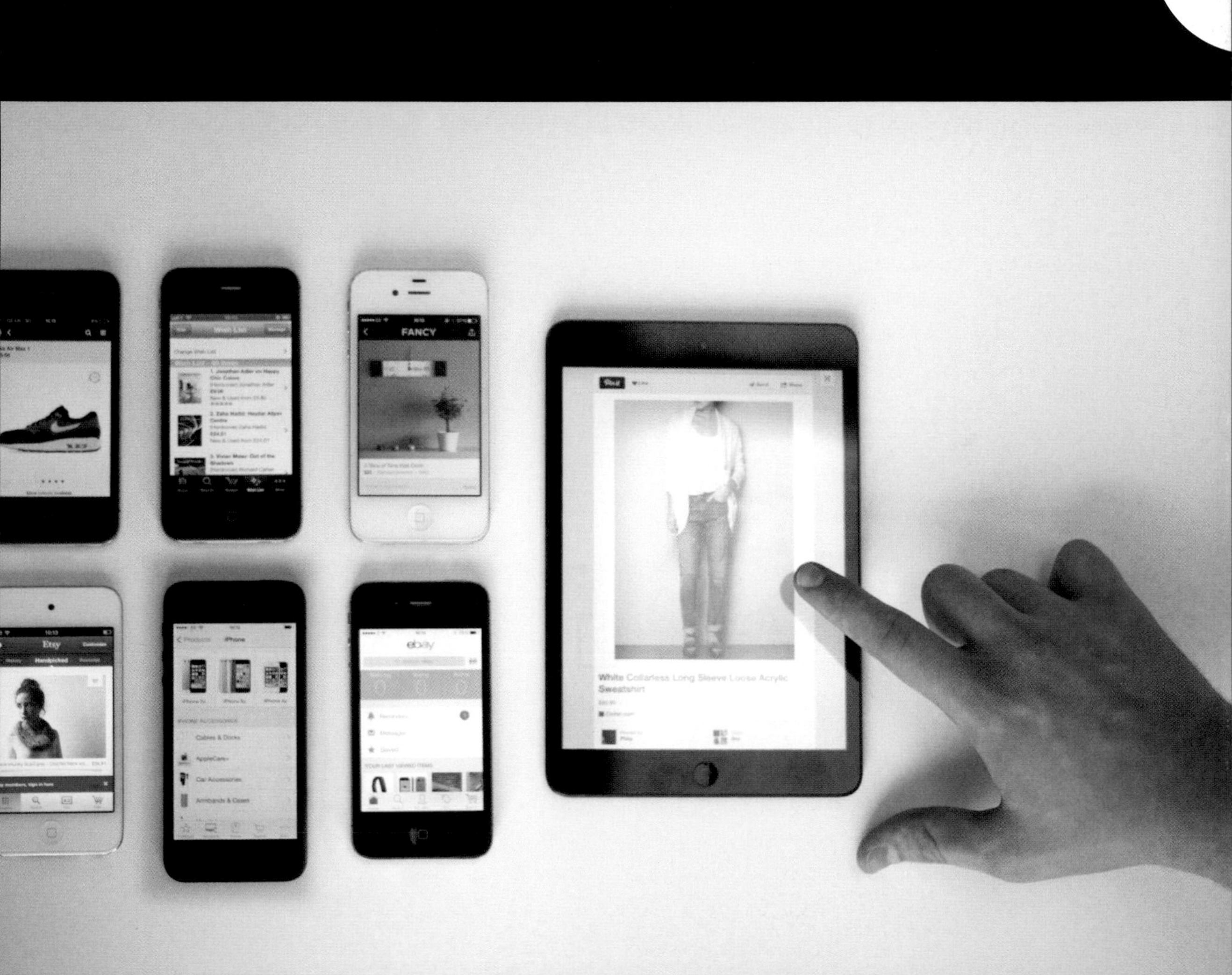
FANCY
Etsy
ebay
White Collarless Long Sleeve Loose Acrylic Sweatshirt

Look at Existing Solutions

As we've already outlined, finding an application with the same premise as your next project needn't leave you downhearted. By researching your opposition you'll find out what's right with an app, but you can also find out what's wrong with it and use that to your advantage.

Though we're dealing exclusively with native iOS apps in this book, you should also be looking at the Android marketplace with your market research, namely Google Play and the Amazon App Store. It's also not a terrible idea to search the net for websites which fulfill a similar function or need as yours. Knowledge is power, and by arming yourself with as much information about your target market as possible, you're making success more of a possibility.

To make sure that you're looking at every possible angle when researching your app concept, it's a good idea to brainstorm all the words you associate with it. If you're stumped, a quick Google search, a thesaurus, or articles and blog posts on the topic of your app (if applicable) might uncover search terms that you might not have thought of. It's these terms that you'll be entering into the various app stores to see if there's an app similar to yours.

It's also worth searching via category. Find the category in the App Store most applicable to your idea and see what's on offer. While an app may not have exactly the same premise as yours, by understanding what's doing well in your chosen categories, you could get inspiration for your design from your discoveries. In fact, at Apposing we recommend that you become an expert on all apps by downloading as many as possible to gain a greater understanding.

Make a note of all the apps similar to yours, always downloading them if possible. While you can get a hint of what's on offer from the app's marketing text, you won't understand how an app really works until you're tapping and swiping your way through it.

If there are absolutely, categorically, and positively no apps like yours on the App Store or net, you've found yourself in one of two boats: either you're on the brink of revolutionizing the app world or your idea is so terrible that no other developer would touch it with a ten foot pole. In this instance you're going to have to rely on opinion, and independent or already existing market research to advise you on how to proceed.

By arming yourself with as much information about your target market as possible, you're making success more of a possibility.

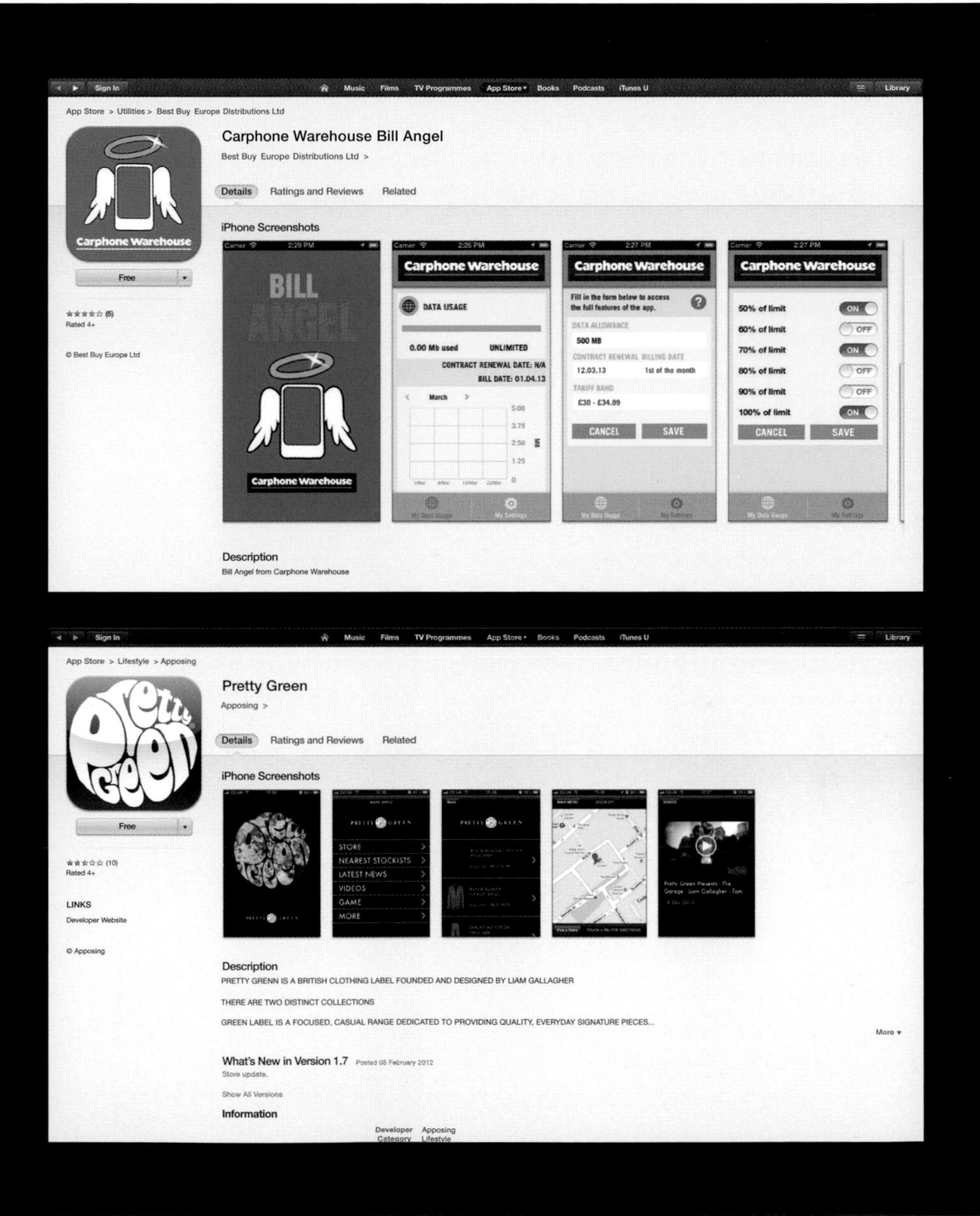

Sign In
Music
Films
TV Programmes
App Store
Books
Podcasts
iTunes U
Library
App Store > Utilities > Best Buy Europe Distributions Ltd
Carphone Warehouse Bill Angel
Best Buy Europe Distributions Ltd >
Details
Ratings and Reviews
Related
iPhone Screenshots
Carphone Warehouse
Free
(5)
Rated 4+
© Best Buy Europe Ltd
BILL ANGEL
DATA USAGE
0.00 Mb used
UNLIMITED
CONTRACT RENEWAL DATE: N/A
BILL DATE: 01.04.13
March
Fill in the form below to access the full features of the app.
DATA ALLOWANCE
500 MB
CONTRACT RENEWAL
BILLING DATE
12.03.13
1st of the month
TARIFF BAND
£30 - £34.99
CANCEL
SAVE
50% of limit
60% of limit
70% of limit
80% of limit
90% of limit
100% of limit
ON
OFF
Description
Bill Angel from Carphone Warehouse
Sign In
Music
Films
TV Programmes
App Store
Books
Podcasts
iTunes U
Library
App Store > Lifestyle > Apposing
Pretty Green
Apposing >
Details
Ratings and Reviews
Related
iPhone Screenshots
Free
(10)
Rated 4+
LINKS
Developer Website
© Apposing
STORE
NEAREST STOCKISTS
LATEST NEWS
VIDEOS
GAME
MORE
Description
PRETTY GRENN IS A BRITISH CLOTHING LABEL FOUNDED AND DESIGNED BY LIAM GALLAGHER
THERE ARE TWO DISTINCT COLLECTIONS
GREEN LABEL IS A FOCUSED, CASUAL RANGE DEDICATED TO PROVIDING QUALITY, EVERYDAY SIGNATURE PIECES...
More
What's New in Version 1.7
Posted 08 February 2012
Store update.
Show All Versions
Information
Developer
Apposing
Category
Lifestyle

Compare Solutions Against Each Other

There's no point knowing what's out there without dissecting it for your own benefit. All of the information that you've gathered at this point should be entered into a spreadsheet or table to allow you to compare and contrast the information with ease. You can download a template that we've created from www.iphoneappdesignmanual.com.

There are a few central areas that we're interested in comparing: the functions, the UX, and visual design. Load time and navigation methods should also be taken into consideration.

If you have multiple competitors, cross referencing the functions that their apps have should make it clear which functions are standard and which are their unique selling points. These factors should be plotted against ease of use—how quickly you can achieve your objective.

While some apps might contain a multitude of features, as we've stressed before, if they hinder the app's primary objective by downgrading the ease of use, then they're adding nothing beneficial to the experience. Take this into consideration as you evaluate each of your competitor's apps.

Mark the features out of ten. Think about whether you could improve on or adopt each of the functions. Use each app as it was intended to get the full experience and see

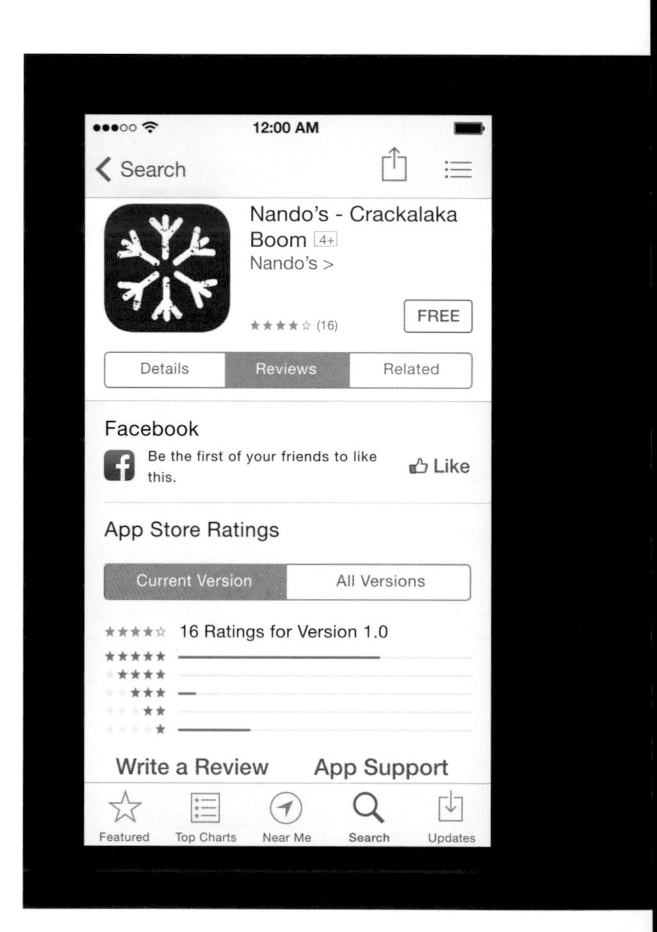

if it's really fit for the purpose it was created for. Remember, bad apps can be as much of an inspiration as good apps. Sometimes, it's not until you see a developer do something badly that you realize how you could make it better. Don't just write it off; figure out why it's bad, why it's hard to use, and you'll have a better idea of how to make yours better.

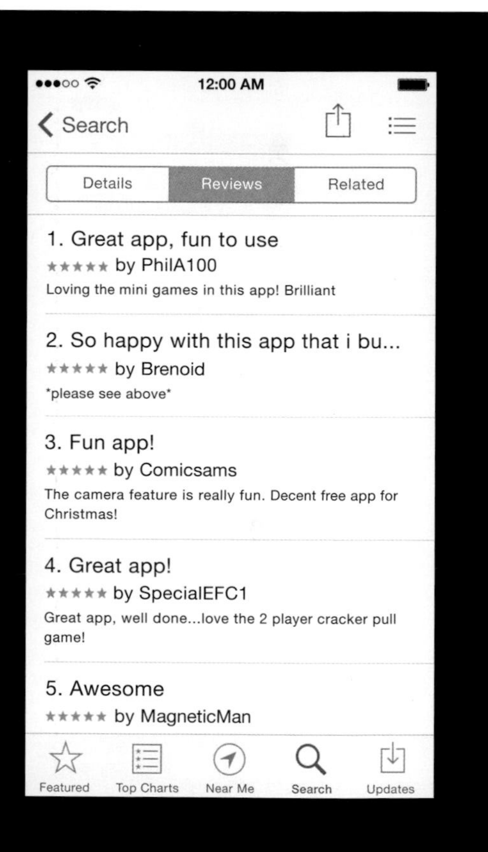

REFINING YOUR APP'S FEATURES

Another thing to factor into your research is the customer reviews—a veritable gold mine of information if the app is popular. People will quickly take to the reviews section when something is wrong with an app, especially if it's paid for. If you've found an app that is similar to yours, the comments section will be filled with your target audience. Use this opportunity to learn how you can avoid the negatives and what your potential customers want from an app of this kind. If you find that there is a common theme with the features that they like, you know to make them prominent. Found a feature that's getting trashed? Cross it off your list. Listening to your customers in this way gives you an opportunity to give them what they really want, and sometimes the opportunity to give it before your competitors can.

Uncover Technical Limitations

There's no point designing a kick-ass app and finding out after development that there was a glaring flaw in your concept that means it's an instant failure. The key to uncovering technical limitations is to discover what's achievable with the feature set you have.

If you're completely clueless when it comes to the feature set of an iPhone, and your research leaves you none the wiser, one option is to look for other apps with that particular function to see if it's possible. Another is to discuss it with a professional.

But it's not just the phone's function list that could hold you back. Imagine you've just designed an amazing app for a music festival featuring a schedule and updates for the revelers. Though a few people will download it before, most people aren't that organized and so would opt to download it at the festival. The trouble is, there's no WI-FI and very poor reception at the festival, leaving those who downloaded in advance unable to receive updates and those who waited until the day of the event with nothing at all.

This stage of the process allows you to explore issues like this beforehand. Map out features that you plan to include, not forgetting your maybes, and outline everything you'll need to make it work.

Also remember that different generations of iPhone and iPod Touch, and iOS, all have differing capabilities. If you want a particular function that will only work on the iPhone 5, and a huge percentage of your target audience have older models, it's worth exploring just how much of your target market will be excluded due to your feature set.

BACKWARDS COMPATIBILITY

The change to iOS 7, though exciting, also posed huge limitations for designers who wanted backwards compatibility. While Apple likes people to progress to the new OS quickly, brands and companies who use their apps to sell services and goods will exclude a proportion of their target market until everyone makes the switch. Therefore, many brands prefer to wait until a larger percentage of iPhone owners are using the latest iOS before making their app compatible.

Plan for Future Functionality

The App Store is a minefield. As the number of apps on offer grows steadily larger, there are a few more tricks you can employ to sustain interest in your app. While many designers spend a huge amount of time designing the perfect app, they forget that loyalty must be nurtured. Though your app might be the top-seller upon launch, how are you going to give continued value and excitement to your customers?

This is where "future functionality" comes in. While matching and even improving on the existing functionality of competitor apps might garner you attention in the short term, it definitely won't be enough to keep you at the top.

You can cultivate allegiance for your app and breed excitement by adding new features in a phased roll-out plan. The phased roll-out has multiple benefits, including increased customer visibility due to the way that iPhone apps are updated. If your app has been going stale on a user's home screen as interest has fizzled out, an update will bring your app back into their consciousness through the Updates section of the App Store. While iOS 7 updates apps automatically, users will still receive a notification that their app has been updated.

The phased roll-out is employed to great effect by Apple themselves. To do it successfully, you need to fathom the ultimate end goal of your completed app and work out a way to release the content in stages, staggering the buzz to keep your audience engaged.

If your app is still held dear in the hearts of your customers, additional functionality, whether free or paid for will add extra benefit for your customers and may lead to them using it more often.

Planning for future functionality will affect all areas of your design. When planning the functions for your app, it's an idea to hold back one or two with a mind to rolling them out further down the line. Remember that adding extra features can affect the navigation and design, so you should factor this into your design process and when thinking about your app's evolution.

For example, Apposing's Bill Angel app for Carphone Warehouse, a leading telecomm-unications retailer, had a phased roll-out plan, where extra features such as a Tariff Band Checker were added after the initial release of the app in the App Store.

Naturally, some desirable functions may come to light when your app is already out there and customers relay their experience with the app to you. While it's great to listen afterwards and react accordingly, getting in there first is always the better option and by targeting and researching your audience correctly you'll be better equipped.

Carphone Warehouse also integrated customer feedback into updates for Bill Angel. Push notifications and alerts for data allowance and contract renewal dates were added after customer consensus deemed them useful.

Future functionality can also come in the form of in-app or "freemium" purchases. A game designer may plan an extra five levels that tie into the central theme, but choose to release them at a later date as an in-app purchase. It's worth looking into the freemium model as it continues to grow in popularity. In-app purchases now count for a huge proportion of the App Store revenue, surpassing the revenue generated by paid-for apps. Still, the model is not for everyone and many developers feel the practice is unethical.

If you build analytics into your app, these could also be used to plan future functionality after release to sustain the life of your app. By studying the patterns of your users through analytics tools such as Flurry and Google Analytics, you can build a better picture of what's working and what's not, adding or adapting functionality according to the information gleaned.

Keeping an eye on technological advances and trends can't hurt either. From iOS updates which will affect design (the transition to iOS 7 took quite a few developers by surprise) and gadgets of note, through to the way in which we use social networks, knowing how we engage and integrate all of these factors into our everyday life is important in UX design. The when, the where, the how, and most importantly, the why, should all be considered when planning for future functionality.

Define Your Target Audience

A common mistake when defining your audience is to assume that they are all like you. After all, you came up with the idea and you find it useful and interesting; why not aim it at like-minded people? Unless your idea is incredibly, incredibly niche, your target audience will spread beyond you and it will affect the UI, the UX, and the marketing extensively.

By knowing who you're aiming your app at, you can execute or get your hands on audience-specific data which will aid you in your planning and design. If that's out of the question, you can use your trusty circle of friends and acquaintances that fit into your target demographic to ensure that you're hitting all the right buttons.

Depending on your app idea, there are various factors which could be important to your target audience:

WHO?

What age do you think that the main users of your app will be? While there will always be anomalies, it's likely that your app will be used by one age group more than another. Age is likely to affect many things: dexterity, income, and accessibility being the primary three. Also, will your app be used by men, women, or both?

WHAT?

Is your app's central theme based on a niche subject matter or interest? If so, your target audience will be infinitely more defined. Find the general topic or category associated with your app's primary purpose and then try to define it further from there.

WHERE?

Where do you expect people to use your app? At a baseball game, in a train station? At home on the couch? Is your app area-specific? The "where" can tell you a lot about your audience and will also affect the navigation and UI.

WHY?

Why will people need to use your app? What circumstances will lead them to pick up their iPhone and open your app? Again, the circumstances used can greatly affect the UX.

Added together, these answers will give you a much clearer picture of who you're aiming the app at.

If you're designing an app for a client's business or your own, then you may have some information already in place on who the target audience for the business is. Use this to your advantage. You may want to target just a proportion of the demographic with your app.

If the app isn't for a business, then a quick trawl of the internet could pull up some juicy stats to start you off with your research. It's the difference between finding out what somebody wants for their birthday present or buying blind. There's a chance you may get it right, but are you willing to take that risk?

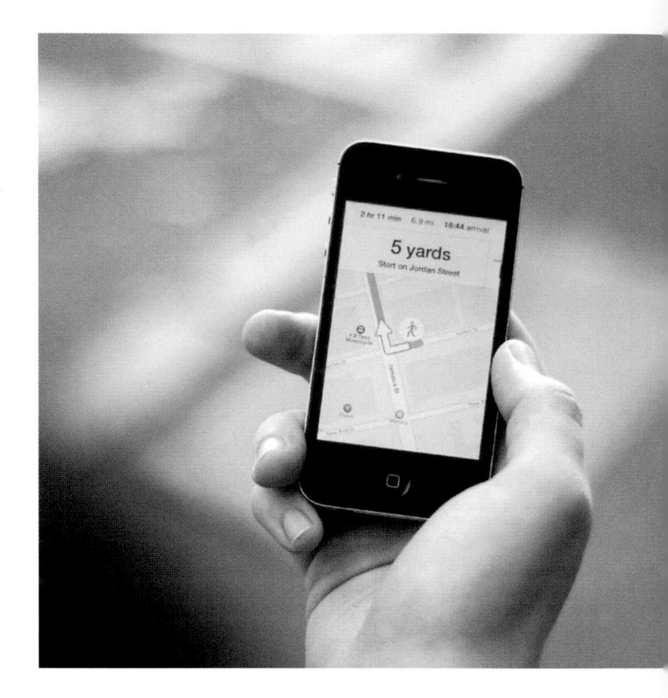

What Do They Want?

Though you have cemented your app's primary purpose, you also need to consider how you want your customers to feel while using it. As we've already stated, people have an emotional attachment with regularly used apps. Apps can give users a sense of achievement, delight, wellbeing, security, and of being in control, among other things.

Popular app Find My iPhone gives its users a sense of security, even when they're not using it, as they know they'll be able to track their iPhone should it disappear without a trace.

My Fitness Pal, another popular app, uses social media integration to add a sense of achievement and also a need to achieve by sharing the user's diet progress and accomplishments with people in their circle. While broadcasting your weight might not seem desirable, announcing you've lost 10lb and having your friends congratulate you might spur you on with your diet.

Taking your app's primary purpose as a starting point, think about how you want your customer to feel after they've completed the function. Adding elements of gamification (such as rewards for tasks), social integration, or support could enamor your app to your target audience by getting under the skin of what they really want. The physical act of swiping away a task on a to-do list is somehow more gratifying than clicking a cross in the corner of a box. Plan functions such as this to add depth to the experience and to leave users feeling how you planned for them too.

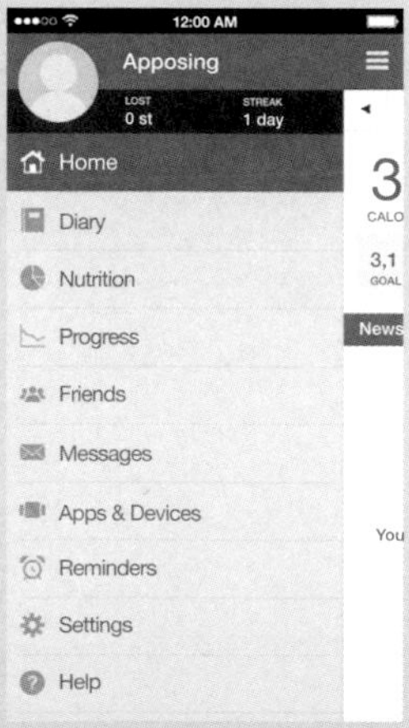

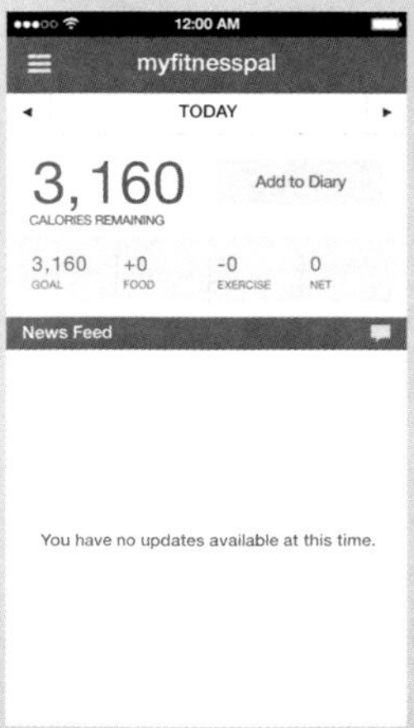

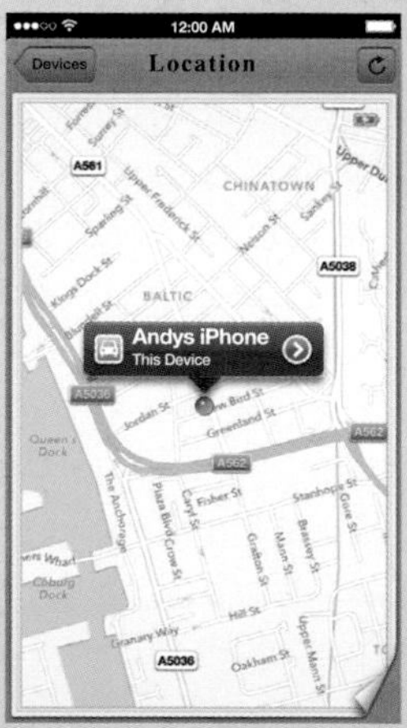

The iPhone Demographic

Do your target audience have iPhones? It's important to find out if the group you're targeting actually use the device you're designing for. And let's say that they are iOS aficionados, which iOS device are they using?

If it's an iPhone, which generation do the majority use, and which OS are they using? iOS 7's launch in September 2013 meant that many people jumped headfirst into the new design and capabilities the OS offers, but it's only supported by fourth-generation iPhones and upwards. By designing to support the latest iOS only, you could be cutting out a proportion of potential customers.

The good news for iOS designers is that Apple customers are quick on the uptake. In June 2013, Apple reported that 93 percent of its U.S. customer base was using iOS 6, just nine months after its release. While other software developers struggle to persuade users to upgrade to a new OS, it seems that Apple customers embrace change more than most. This bodes well for app designers wishing to stay ahead of the trend by designing for the latest iOS without cutting out a huge proportion of their potential customer base.

iOS 7 was the biggest shake-up of iOS design since its conception. It moved away from the skeuomorph; design became flat; white space and transparency were in; and it brought a host of exciting new options for developers.

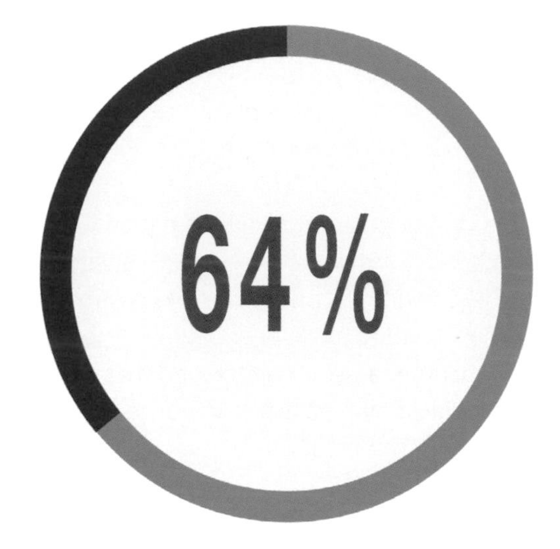

Switch to iOS7

Developers faced the option of designing for iOS 7 outright, and waiting for the world to catch up, or designing apps which suited the iOS 7 aesthetic but still supported an older iOS. Some developers would have even gone as far as to design two UIs, but this is not advisable as it dilutes your brand through an inconsistent look across multiple platforms.

ComScore research (Feb 2013) shows that, on average, 69 percent of iPod Touch users are aged between 13–24 years old, a much younger bracket than the iPhone.

iPads remain popular with all the family, though the results skewed toward the male owner (52.9%), and slightly younger than the iPhone (44.5% under the age of 35). It's also worth noting that the same research found that 46.3 percent of iPad owners reside in households with an income of $100k or greater. Though these are merely stats, they could form the basis of further research to establish whether you'll be hitting the right market with your efforts.

Statistics on who is using each device are fairly easy to track down online. Try to find stats from independent parties; this way you won't be looking at a clever marketing ploy. Remember, when you make an app for an iPhone, you'll be making one for an iPod Touch automatically. iPads, however, have a very different screen size and so need to be handled separately. You'll need to make a decision to develop your app for the iPad as well.

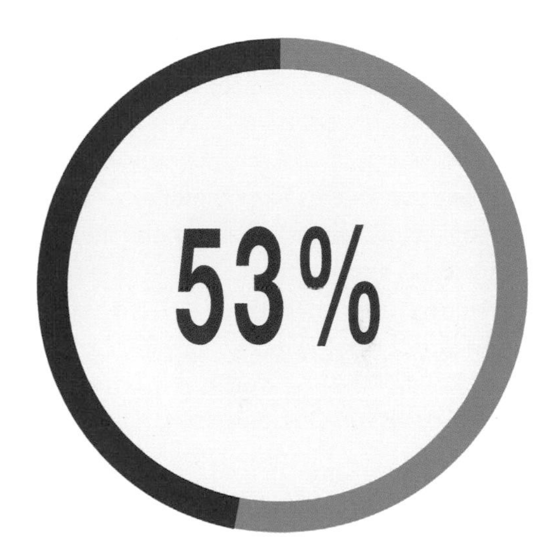

iPads owned by men

How Your Target Audience Affects Design

In terms of design and later in the process, marketing, knowing your audience is invaluable. From the App Store icon and screenshots to the price and marketing copy, you can't tailor them to maximize downloads if you have no idea who you're tailoring to. Big app developers spend lots of time testing everything from different navigation set-ups to minute details such as icon color to see which resonates best with their target audience.

There are other factors to consider too. As outlined before, age plays a huge part in how the UI and UX are designed. Young children are less dexterous and less inclined to learn how to use intricate control systems. They like bright, primary colors. They like music, characters, and sound effects. Even when learning, they like to learn through play. All these factors will affect the UI and UX. But while it's important to recognize who will be using the app, it's also important to think about who'll be buying it. Many parents now let their children use their phone to keep them occupied. So while a children's app must hold the attention of the child, the marketing materials (marketing text, icon, and screenshots) must also appeal to the parent, as it is the parent who will part with their cash.

There's been much controversy recently about the use of in-app purchases in children's apps. Freemium apps (meaning free apps which offer the user premium content) often employ psychologists to help them design a game that pushes the user into buying premium content, either by making them wait or allowing premium content purchases to allow you to progress further. Parents are wary of such things, and prefer to pay once and once alone. The uproar led to Apple highlighting which apps contain in-app purchases in the App Store. Keep an eye on topical news regarding your app as well as the industry to see what the general consensus is saying.

If you're designing an app for a client, you'll almost certainly have brand guidelines to adhere to, as well as tailoring the app to the target audience.

Try to find out everything you can about your target audience, from marital status to their average income. Their hobbies, the apps they use currently, whether they pay or stick to free apps, are all factors that can influence your design. Whether you find hard stats, ask friends, or poll your social media followers, any information is good information.

16:28
95%
Search or enter an address
A580
LIVERPOOL
ROPE WALKS
WATERFRONT
CHINATOWN
Hope St
BALTIC TRIANGLE

CHAPTER FIVE

INTERACTION DESIGN

It's important not to think of design and development as entirely different processes. If you haven't planned who will be developing your idea, now is the time to do so. iOS 7 has a greater lean towards interactivity and gestures than iOS 6, and by launching headfirst into your design process you might be missing out on interactivity, or worse, planning for tasks that simply can't be accomplished.

The line between designer and developer has become increasingly blurred. Both parties need to work together to achieve something that does the iOS 7 operating system and its capabilities justice. You should talk to your programmer about gestures, OS iterations, transitions, animations, and about other user-experience functions before you fully commit to the design process, to avoid work being wasted.

There are plenty of resources on the web that can guide you when choosing a developer if you're not going to be doing the work yourself. Though this book won't detail the hiring process, we will be identifying key areas and points that you can talk through with your developer when you've chosen.

What is Interaction Design?

App interaction design is a deep understanding of how a user interacts with a piece of software put into practice. Focused on usability and user behavior, good interaction design solves problems the user might have in accessing the required information when performing a task. It also creates a smooth and fluid method of extracting the right information or initiating the right action from a user to be able to use your app.

It's swiping to access your phone, it's pinching to make a screen smaller, it's tilting your screen to change to landscape view, and shaking it to undo a command. It's what makes using a touchscreen phone so immersive.

In terms of an application, it's using the body, mainly your hands and touch, to create a dialog with your idea. Instead of information being foisted on your end user, the information will be presented in such a way that the user will interact with it, creating a deeper virtual discourse with the person using your app. The user will tap, swipe, shake, and slide their way to success. The iPhone offers designers a vehicle in which the phone can become pretty much anything you need it to be. Functions such as the light, accelerometer, vibration, compass, and the inclinometer allow the hardware to become anything from a torch to a spirit level, a steering wheel to a remote control.

With each new OS, there are a multitude of new functions that appear for designers to exploit. For instance, iOS 7 introduced the inclinometer, a feature which measures the angle of slope/tilt with regards to gravity.

The eagle-eyed among you will have already picked up on the fact that apps that performed this function preceded the introduction of iOS 7, and you'd be right. The introduction of the new OS was just a software update; nothing actually changed in the phone's hardware. The ability to measure angles and gradients has always been in the phone, and has already been used in applications such as Clinometer, a bubble level and slope measurement tool currently available in the App Store. What has changed, however, is that the OS update brought with it a new programming interface, or API, which allows all developers to use the function in their apps.

Developers use APIs to utilize the phone's hardware in a particular way. Knowing the phone's software capabilities as they stand will help you to formulate a stronger app, but talking your ideas through with your developer might uncover new options that you've previously thought impossible.

The same applies to gestures. In the iPhone's infancy, the now widely used pull-to-refresh function wasn't one of Apple's controls. It was created by a developer called Loren Brichter,

Just because Apple haven't created an API for what you want to do doesn't mean it can't be done.

ALWAYS THINK BIG.

who made the gesture for the app Tweetie, which is now owned by Twitter. If you have a particular interaction in mind that doesn't currently exist, talk to your developer. He or she can advise you on whether you're simply trying to reinvent the wheel or if a gesture already exists that could fulfill your function. Alternatively, your developer might feel that your new gesture is worth taking a risk on. Creating an app is a two-way dialog. Don't simply see your developer as someone who can code your design; instead value their experience and knowledge to add value and innovation to your idea. The best ideas do something different, using objects we've come to take for granted in a new and exciting way. Your developer is the key to making your dreams for your app a possibility.

Don't just think about creating a piece of software, think about creating an experience for your users. The iPhone is very versatile; let your imagination and understanding of your audience shape what you create instead of restricting yourself by thinking of its functions in limited terms.

THE iPHONE'S FUNCTIONS

There are plenty of hardware functions on offer for you to take advantage of. These can add an extra dimension to what you create.

- Front-facing camera
- Rear-facing camera
- Video recording
- High resolution, multi-touch display
- GPS
- Gyroscope
- WI-FI & 3G internet
- Bluetooth
- Speaker
- Microphone
- Headphone jack

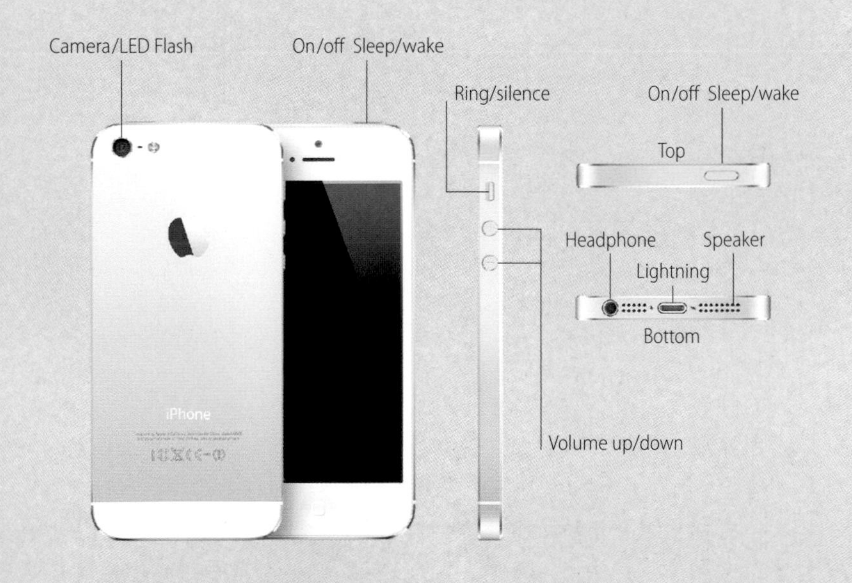

Designing for Gestures

Ultimately, you want your users to be experiencing your app, not the gestures. Complex is never better. Apple have already pretty much covered 99 percent of all the gestures that you'll need, so it would be crazy to start designing new gestures that fulfill the same function as Apple's own controls. People are well used to the Apple controls, and it would be counterintuitive to ask your users to learn new ways of doing something simple.

The majority of objectives can be achieved by tapping and swiping your way through the information, though there are other, less heavily used gestures that can make your app experience more fluid.

Planning interactions into your design is part of the process that turns an app from just another piece of software into a personal experience for the user. Choosing the right interactions allows people to make using your app a personal experience as they touch their way through it instinctively.

There are other well-used gestures, such as pull-to-refresh. Pull-to-refresh is now patented by Twitter, though it's a very loose patent and you should feel free to use it. The flick is another great gesture. It allows you to move content as you would with a swipe, but much quicker.

When thinking about which of these to use in your app, try to think about how people go about things in the real world. While we don't mean to keep mentioning Loren Brichter (he's considered a bit of an interaction god in the app world), his thoughts on creating interaction are sound. He thinks up new features and interactions for apps through watching how people move objects and complete tasks in the real world. "Everything should come from somewhere and go somewhere," he explained to *The Wall Street Journal* in 2013. Annoyed with apps that feature unrealistic interactions, such as pop up or collapsing menus, he explained, "the most important thing is obviousness," with the problem being overdesign.

With this in mind, study how people interact with things related to your app, and try to mimic the actions in your interactions if possible. Of course, some people develop habits which are more time-consuming than they need to be—don't mimic those if there's a quicker way. The key is to use the right gestures for your idea, and to do that you need to tap into the mindset of your target audience.

APPLE'S CUSTOM GESTURES ARE:

Tap: Use to press or select a control or item.

Drag: Use this gesture to scroll up or down a page, or pan from side to side. It can also be used to drag an element.

Swipe: This can be used in a number of ways. In a table, swiping a piece of information will often reveal the ability to delete that cell. Swiping from the top of the screen will reveal the user's notifications, and on iOS 7, swiping from the bottom will reveal the control center. A horizontal swipe moves content onto or off the screen. The new OS also allows the designer to show more or fewer options by recognizing a "hard" or "soft" swipe. A soft swipe brings up the regular options in the iOS 7 Mail app, such as delete a message, but a hard swipe will bring up other options available to the user.

Double-tap: Use this to make a piece of content, such as the text, full screen and centered. It can also be used to zoom in or out. Double-tapping also selects editable text to allow copying, cutting, or other functions.

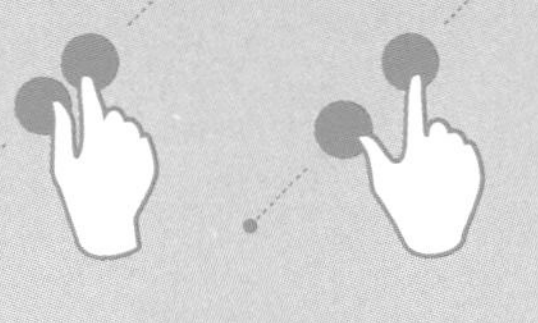

Pinch: Pinching the screen open zooms in, while pinching together zooms out.

Touch and hold: Use this in editable or selectable text to display a magnified view for cursor positioning.

Shake: Commonly used to initiate an "undo" or "redo" action, shake is used for many different things in apps across the App Store. In the Music app, you use the shake gesture to change songs when in shuffle mode.

Pan: Tapping and holding on an object to select it, then dragging the image around is a pan gesture. For example, moving around a panoramic picture in zoom uses the pan gesture.

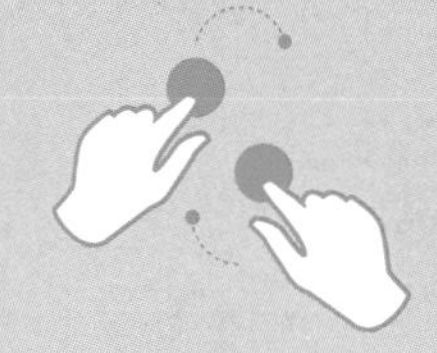

Rotation: Much like the pan gesture, rotate requires you to tap and hold an object such as an image, but this time, you use a second finger to rotate it. You can move a map round using rotate to get a different view.

Long Press: A long press or tap-and-hold gesture actions many shortcuts on the iPhone. For instance, a long press on an image brings up a view which allows you to copy or save the image. It also reveals data that's used less regularly, such as accented letters on a keyboard.

The Accelerometer, Gyroscope, Compass, and Inclinometer

Gestures are only half the interactivity battle for apps. An iPhone has a whole host of other features that allow your device to become whatever it needs to be. These hardware features are listed in the diagram shown earlier in the chapter on page 69.

One hardware feature you should consider using is the accelerometer. The use of the accelerometer in apps has grown substantially in the past few years, as developers have looked to movement to create a more immersive experience. On a basic level, the accelerometer can detect whether you're holding your phone in portrait or landscape mode, and adjust the display accordingly. It's also the hardware that allows you to tilt games to control them, or to shake your phone to start a function.

More specifically, there are sensors in your phone that measure both movement direction and movement speed on three axes; right to left, up and down, and back to front. The accelerometer opens up a lot of possibilities for interaction design.

You'll already likely be able to think of various gaming apps which utilize the accelerometer, such as the smash hit Temple Run. The game requires the player to tilt the phone to move the character up the level.

While the accelerometer on its own is a powerful piece of hardware to have on your side, there have been other new additions to the iPhone setup which allow you to further make use of movement.

The iPhone 4 saw the introduction of the gyroscope. This hardware addition was big news for developers, especially the gaming community, because it allows for greater control of the phone. Whereas before you could move the phone to turn content forward and backward and from side to side, the accelerometer didn't register movement precisely, as it couldn't register physically turning the phone panoramically. The built in gyroscope adds the ability to measure another three axes to the setup. This means you can measure movement on six axes. If you keep your feet in the same place and turn 360 degrees, the phone will use the gyroscope hardware to recognize that movement.

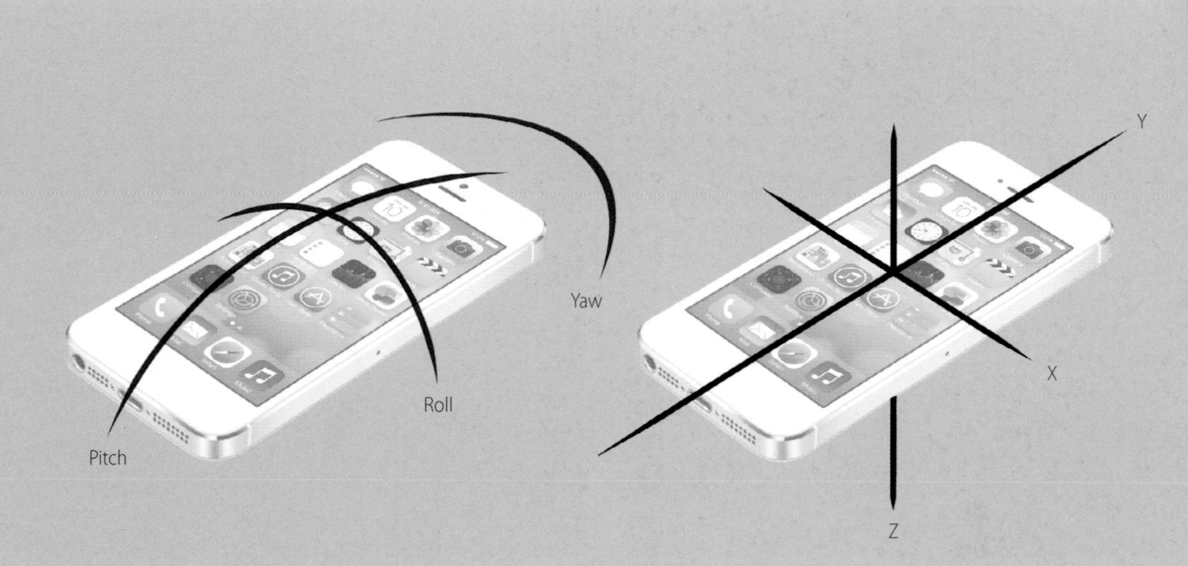

Mathematically speaking, the accelerometer measures movement and speed of movement on a three-point axis; the gyroscope takes the movement axis up to six, so that the movement that appears on screen can be panoramic, or even omnidirectional. This is great for apps which require a 3D functionality or rotation, and increases the number of slight gestures that you can design for.

One great use of the gyroscope in iOS 7 is the moving wallpaper on the home screen. If you upload a panoramic picture as your home screen background and turn panoramically, the background image will move with you, leaving you able to see the whole picture by simply turning. This effect, where the foreground stays the same but the background moves to suggest motion, is called a parallax. The iOS 7 parallax also has an effect on still images, making it seem as though they float above the background. This effect is achieved by building your screens in layers. The gyroscope then moves each layer individually, attuning the movement to your motions and speed. The layers at the back will move the quickest, and at the front the slowest, creating the effect. This effect has been used since the gyroscope's introduction in 2010 in apps which create hologram effects.

We used the accelerometer to great advantage in our Christmas app for the restaurant Nando's, titled Crackalaka Boom. The app was a virtual one- or two-player Christmas cracker with a host of goodies inside. We created the feel of a traditional Christmas cracker by using the accelerometer. Using Bluetooth technology to allow two phones to connect with each other, the two-player users held their phones together as you would a cracker. We then used the accelerometer to measure the pulling speed and movement as the users "pulled" the cracker.

You can measure with some degree of accuracy, where and how fast the user moves the phone, even if they turn around on the spot.

The gyroscope, coupled with the accelerometer, can make for better augmented reality apps, i.e. apps which create an alternate reality on the screen by combining software and real life. The addition of the gyroscope means that the user's movements are more accurately represented on the screen, as each tiny movement is now measured on six axes.

When iOS 7 was launched, Apple announced that there would now be readily accessible software support for developers wishing to use the compass and inclinometer. As we mentioned earlier, many iOS 6 apps have made good use of the inclinometer by creating their own APIs to use the hardware. The inclinometer measures the tilt or angle of the phone in relation to gravity. When used with the accelerometer and gyroscope, you can combine rotational and horizontal movements with vertical movements, for example, to measure climbing stairs. It's likely that there will be an increased number of apps that use this extra dimension of 3D space tracking technology.

The iPhone's inclusion of all three pieces of hardware means that you have a world of different control mechanisms that are now at your disposal. This is great news for interaction designers, as the tools at your disposal have never been more versatile. From basic apps that use the accelerometer to silence a call by simply turning your phone face down, to games that combine all three to create a hyper-real experience, what you can create is really only hampered by your imagination.

The M7 Co-processor

The launch of the iPhone 5s added a new level of movement measurement for interaction designers to take advantage of, with the introduction of the M7 co-processor.

This exciting addition to the iPhone family opened up the realms of what you can do with your phone, with a three-axis accelerometer, a three-axis gyroscope, and a three-axis electronic compass. "Nothing new here," we hear you cry, but the novel part is that it measures motion data from all axes continuously, even when the phone is asleep, and without draining the battery as heavily.

While iPhones have long contained these motion sensors, they are turned off when the phone is asleep to avoid overloading the main processor and draining the battery. The M7 operates separately from the main processor, and combines and stores all the information from each individual sensor. Apps are then able to access this wealth of historical and current information, so a newly downloaded pedometer app will also be able to backdate your results from the information stored in the M7.

As the M7 co-processor is always analyzing data, it gives a more rounded indication of a user's movement, i.e. it doesn't just measure how many steps you're taking when the phone is on or when you're using a particular app, it also measures how many steps you're taking when the phone is asleep in your bag. It also recognizes when you're in a car or running. Apple used the M7 co-processor to solve an annoying problem found by iPhone users driving in their car: unless you disabled the setting before you started driving, the phone would ask to join WI-FI networks it picked up along the entirety of your journey. As the M7 co-processor can tell when you're walking, running, or driving, this setting is now disabled when the phone recognizes that it is in a moving vehicle.

Fitness apps have been the first to employ the M7 co-processor to great advantage. Apps such as Nike+ Move and Argus were quick to publish their apps that utilize the new processor as soon as the iPhone 5s was launched. Users downloading these apps after getting their hands on a 5s were able to see data ranging from the day that they got their phone, not the date they got the app, due to the way that the processor works.

On a basic level, the M7 can do everything we explained in the previous section; the accelerometer, gyroscope, compass, and inclinometer measure movement in exactly the same way in the 5s as they do in the iPhone 5 and below. The takeaway here is that measuring these movements will have less impact on the battery life of your user's phone, and most importantly, the processor captures and stores movement information even when the phone is sleeping. The information stored in the processor will allow you to populate an app with data that precedes the download date.

FINGERPRINT SENSOR

The iPhone 5s also saw the introduction of Touch ID, a fingerprint sensor in the home button that bypasses the need for the user to enter a passcode when unlocking the phone or making purchases in the App Store. iOS developers naturally became excited, detailing the many different ways that they could use the technology. At the time of writing however, Apple were quick to pour water on that fire, announcing that for the time being, the fingerprint technology won't be open to developers. So if you've been planning an app that uses the fingerprint scanner as an integral function, it's time to rethink that notion for now . . .

Though the advent of a new processor might not seem to matter too much to an app designer, as the introduction of the M7 co-processor shows, it matters a lot. While you can leave understanding how the hardware works to the developers, it's important that you make note of the new possibilities that the tech brings to the table. The M7 has revolutionized the way that fitness apps work forever, and in time, the processor's abilities will drip down to other app genres. For each new piece of hardware that a new phone model brings, make sure that you note down what it does, and brainstorm if these new introductions would have a positive impact on your app idea for future updates.

How to Use Other Software/ Hardware Features

There are another raft of hardware functions available to help turn your concept from an idea into tactile and intuitive app. Though some of them will seem obvious, forgetting about them could mean you miss out on really useful possible functions.

For many people, their mobile has replaced many household and personal items, including the wristwatch and alarm clock. Not charging your phone overnight can seem catastrophic come morning, as you're left without a wake-up method and way of telling what time it is. Synched calendars mean that it's easier to be alerted to upcoming obligations and events. How did we live without our iPhones?

Though one of the more basic functions on the iPhone, the clock offers many great benefits for designers. For example, the Facebook app utilizes the clock function to great effect on their business pages. Each business enters their opening hours upon creating the page, and when a user looks up the business on Facebook, they are alerted as to whether the business is open or closed. If it's closed, the app lets you know when it will be next open. This is invaluable information for today's consumers who are short on time and patience. Instead of trekking over to a shop only to find it's closed, users can check on the move whether or not it would be a wasted journey.

If your app idea is productivity based, the clock and calendar will no doubt prove invaluable. For those of us trying to organize our lives, the ability to set reminders and alerts can make or break this kind of app. Super basic apps such as birthday reminder applications have saved face and feelings on many occasions. Instead of just reminding someone on the actual birthday, you're given the option of a reminder days or weeks before so that you have ample time to buy a present and card; much better than being left red-faced and present-less on the day.

Though the clock and calendar aren't the sexiest functions you'll find on an iPhone, and are the least likely to produce innovative applications that beggar belief, these two are the bread and butter of the function world, impossible for most users to live without.

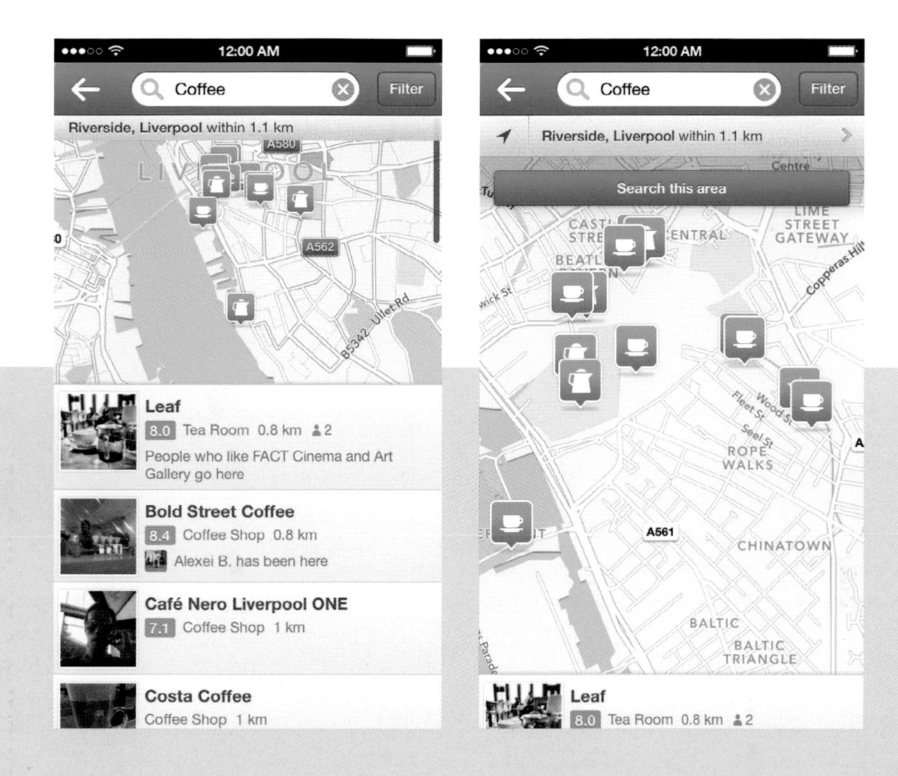

GPS

It's difficult to remember what we did before we had a map of the world in our pockets. It's rare to find yourself in a zone which isn't covered by a map of some description. But GPS is used for much more than last-minute directions to your job interview.

Finding your friends at a festival (providing their phone hasn't already run out of battery!) is now a cinch thanks to clever use of GPS. The sports industry has taken advantage of GPS to improve training. Coupled with the accelerometer, apps such as Nike's running app were early adopters, creating an app which correctly measured route, and by incorporating the clock, was able to accurately assess the speed at which the user covered the distance. Yelp and Foursquare are also great examples of well-used GPS. Foursquare managed to gamify the GPS function by giving awards to the person who checked into establishments most regularly.

The GPS function is also widely used to tailor search results according to where you are. If you enter a search for coffee into an app, after locating your position, the app can pull up the closest shops to you, with an estimation of how long it will take you to get there (and of course, using the clock function, tell you whether the shop is open or not!).

CAMERA

This function has been used to best advantage by apps such as Instagram and Vine. The multitude of apps that take advantage of the user's ability to take a photograph at any time in any place is staggering.

As well as allowing users to put filters over photographs as in Instagram, there have been many successful apps that allow you to edit pictures once taken. There are apps that allow you to collage the photographs that you take, swap the faces in a photograph, and even apps that allow you to create short stop-animation films or "cinemagrams" and gifs. Vine, a social network in its relative infancy, allows people to create easily sharable 15-second videos that can be quickly put together with no external editing equipment.

Don't write off the camera as a menial and supplementary addition to your app. It's not just a method of allowing people to take selfies to post to Facebook.

The iPhone possesses a powerful rear-facing camera, and a less powerful front-facing camera. It's the rear-facing camera which is used more often than not for apps that take pictures or videos, and the front-facing camera which is used for communication, such as in Apple's Face Time, and more recently, for accessibility interaction controls that allow you to make gestures with your eyes and head to control an app. We've been fascinated by these, and believe that these gestures will soon become commonplace in games (how about a reload from simply tilting your head to the side?) and other everyday apps.

Translation apps use the camera and text recognition to help travelers find their way about and make choices by translating signs, menus, and more by holding the camera over them.

We've used the camera function many times in our applications. We used it in a fairly traditional way in our Nando's application, Crackalaka Boom, to create a Christmas-themed photo booth, in which users could add festive hats, earrings, rosy cheeks, and scarves to pictures of themselves.

We made great use of it in our award-winning augmented reality application for sofa retailers CSL. Instead of having to trek to their nearest CSL store to view a sofa and still have no real idea of how the sofa would look at home, we used the camera to allow people to superimpose their chosen sofa into their living room, giving them a better idea of what they were purchasing.

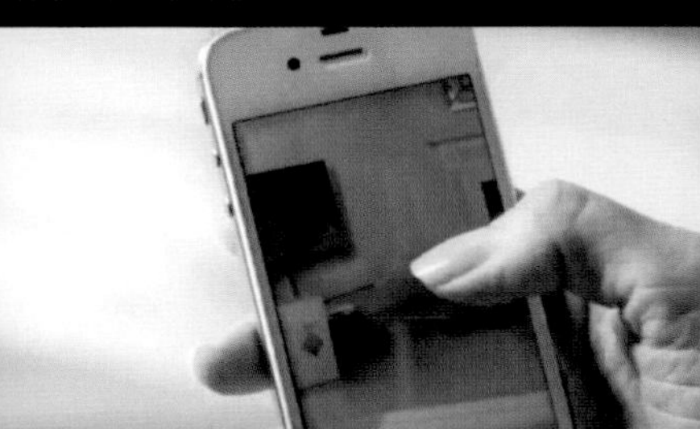

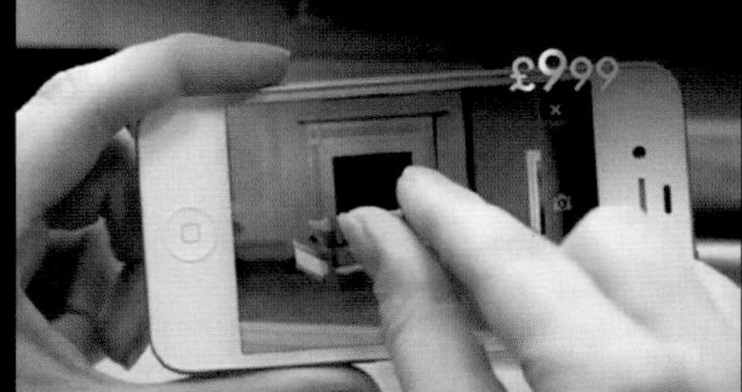

VIBRATION

The vibrate function will no doubt strike you as an almost silent mode of alerting the user to an event of some description. Most of us will be well acquainted with a short vibrate alert when we receive an email with our phone on silent, or a longer vibrate for a silent call. Games often employ the vibrate function when the player strikes a wall or hits a target, or to alert the user that they're doing something wrong.

Alert function aside, the vibrate function can be used to great effect to create a more immersive experience for your user. Some apps use the vibrate function, caused by a motor running inside the iPhone, to amplify effects or sounds. Generally speaking, app developers use the vibrate function to communicate with an app's user when sound would be a distraction or a problem. When thinking about the many situations in which someone will use your app, you should also think about whether or not sounds caused by your app will be a problem. If they might be, will the usability of your app be lessened if the user puts their phone on silent? If yes, you could consider using vibrations to replace sounds. You can even create custom vibration patterns for calls or alerts so that the user knows who is calling or what the alerts are for by vibration alone.

The app really demonstrates how thinking outside the box can take a seemingly menial function and use it to great advantage for the user.

People have started using the vibrate motor in more creative ways, with apps such as Cycloramic taking full advantage of its capabilities. Used for taking 360-degree panorama images and videos, the app aimed to solve the problem experienced by many whose hands were just too shaky to take a stable 360 picture. Enter the Cycloramic app: it allows people with access to a flat surface to take perfectly steady pictures by making the phone vibrate at just the right frequency to make it twirl steadily, continuing until the user presses stop.

12:00 AM

What can I help you with?

AUDIO INPUT/OUTPUT

We're all well versed in using the iPhone's audio capabilities to make calls. The audio input and output predates the smartphone, but the age of the iPhone has seen people use the hardware to create an endless supply of bright and brilliant apps.

On an obvious note, the audio functions are used in video and music apps, podcasts, iPhone alerts and alarms, and to add an extra dimension to gaming through music and sound effects. Even very visually simple apps such as Apple's Mail app use audio to alert the user that an email has been sent or received. Sometimes, utility-based apps utilize audio to keep users on the right track without having to crowd the interface with visual alerts.

Sound is often a neglected medium in apps. You should try to use it sparingly and to your advantage. Anyone with the Twitter or Skype apps will instantly recognize their sound effects. This extends their branding beyond the visual. Overcrowding your app with sound effects will have the opposite effect, and may lead to the user turning the sound off completely. Use sound to supplement your user interface and to create a better user experience.

It's important to think about how much a sound or sequence of sounds will affect the usability of an application. If you decide to use a sound to alert a user that a task has started, what happens if the app is being used in a place where the sound will need to be turned off? Make sure that you consider that many people have their phones on silent much of the time, and factor this in to your design so that usability is still high with no sound.

One of the most exciting additions to the iPhone was that of Siri, a built-in personal assistant app which uses voice recognition technology to allow you to ask questions, give commands like scheduling meetings, and more. The app can then ask the user questions to gain more information to be able to answer the question, or complete the command, more efficiently. The 4S also saw greater accessibility options appear on the keyboard, leaving users able to speak into their phone instead of typing. Appearing as a microphone button on the keyboard, users can dictate notes or their personal data into forms. It's worth noting that the API for this is not widely accessible; if you're planning on creating a speech recognition app or something similar, you'll need your developer to create their own API.

Other apps of note which use the audio functions well include one of the most successful apps to ever appear on the App Store, the Ocarina. This app turns the iPhone into an ancient flute by allowing the user to blow into the microphone. The user then uses touch functions to change the note. If you've ever used this app, or other games that require you to blow to fulfill a function, you'll know that it's a fun and novel way to interact with your phone. That being said, it's fun every once in a while. If you were creating an email app that required you to blow to delete a message, despite being fun at first, you might feel a little embarrassed blowing your way through an influx of mail and would eventually end up hating it.

Many apps, such as the Google app, grant users the ability to speak or type a request. Remember that you might not always be in a situation in which you can speak your request, so if voice controls will add an extra element to your app, remember to also leave the user the option to type their request in case saying it aloud would not be acceptable.

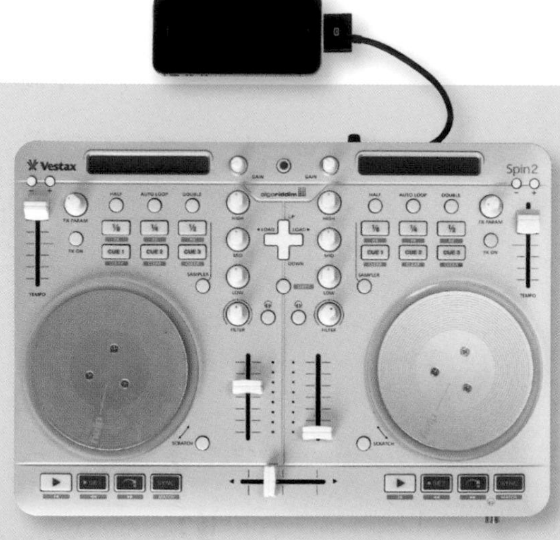

Mule's Ocarina

Mule's Ocarina

Many developers connect external hardware to the iPhone through the audio input rather than the lightning connector or charging port on the bottom of their phone. There is a reason for this. There is a rather hefty developer charge for using the charging port as a connector in your app, so many developers avoid paying it by connecting via audio input instead. We used the audio input to connect a breathalyzer to the phone a few years ago. Using Arduino, an open-source electronic prototyping platform, which can sense environmental changes, the user was able to breathe into a sensor, and the app used the information gathered to ascertain whether they were over the limit.

Playrise's Table Top Racing

BLUETOOTH

Here at Apposing, our earliest memories of Bluetooth revolve around connecting to random, unlocked phones in crowded places, and sending messages to oddly named phones to see if we could spot whose device it was. Thankfully, things have moved on a lot since then.

Bluetooth, a form of data exchange using shortwave radio signals, is widely used for wireless connectivity to keyboards and mice, headphones, and mobile hands-free kits in cars. The technology is also used to connect phones for multiplayer games, chat, and even to make calls. It has also become extremely useful for file sharing.

As we mentioned, we used Bluetooth connectivity in our app for Nando's to temporarily connect two phones to recreate a cracker pull. For games or fun apps, the Bluetooth function can be used to recreate real-world movements with another person. Games such as Table Top Racing allow the user to connect via Bluetooth to play multiplayer games.

Bluetooth is widely used for wireless connectivity to headphones and mobile hands-free kits, and is also used to connect phones for multiplayer games.

WI-FI AND 3G

This one is so obvious that it pains us to say it, but it brings up a few considerations you should factor into your decision-making process.

Though not necessarily linked to interaction design, like Bluetooth, many apps such as Liveview use a WI-FI connection to transfer data from your Mac to an iPhone or iPad. It's a useful tool for app designers as it allows the designer to view their UI on the device of their choosing, even allowing them to test out their interaction design with software prototypes.

We bring up WI-FI as it's another method of connecting your user with other devices. Social interaction is also a major part of the app boom. Do you want to factor social connectivity into your idea?

AIRPLAY

AirPlay is an Apple software component, introduced in iOS 7, which allows for wireless streaming of video, audio, and photographs. The data can be streamed to a number of devices, including HDTV and Apple speakers. Bluetooth can be used to stream music, but you'll need a WI-FI connection for photo and video.

12:00 AM

Cancel Printer Options

Select Printer >

1 Copy – +

Print

AIRPRINT

AirPrint is pretty cool as long as you have access to an AirPrint printer. It allows instant printing from iOS devices without the need to install drivers. Just connect to the printer via WI-FI, and you're good to go.

We've only touched on a few different uses for these hardware functions, and this is where you come in, finding ways to further your idea by bending the technology to your advantage. We've added a full list of new features introduced in iOS 6 and iOS 7 at www.iphoneappdesignmanual.com.

Don't be pessimistic when planning your interaction design. Aim high; the chances are there's a way or at least a workaround available to you.

CHAPTER SIX

USER EXPERIENCE DESIGN

The previous chapters should have led you to a point where ideas are spilling from your head. With a firmer understanding of what you can do, this chapter will help you to get those ideas in order. This section deals with user experience design, or UX design as it's widely known. We'll be taking you through wireframing, case scenarios, and will tackle the flow of your app. We'll also take you through the various transitions and navigation styles and lock down how the information can be presented. This chapter is a big one in which we explore many concepts. Have a pen and paper ready to note down the bits that are relevant to you.

What is User Experience Design?

As a consumer, you shouldn't really be aware of the user experience. Done correctly, the hours of toil, experimentation, and frustration put into it become invisible, evolving into an intuitive and fluid experience which allows the user to complete a task or function in the easiest way possible. You only really notice the user experience when it's executed poorly. Quite simply, user experience is the art of making things work in the best way possible for your target audience. To do this, you need to know how people react and interact so that you can make decisions based on their behavior.

This chapter will incorporate what we learned about interaction design. Your user interface, or UI (which we explore in the next chapter) will be built upon your UX framework.

It will also make you think long and hard about the way you approach delivering your app's primary purpose and how people move about your app. User experience isn't some new-fangled, faddy buzz term. Done considerately, it could change the way that people perform a task forever.

There are many factors that can contribute to iOS UX: the flow of the app and the navigation methods used, the transitions (i.e. how you

> ***It's up to you as the designer to create something that leaves the user feeling like they've achieved their objective. Good UX is your idea coupled with the correct implementation.***

move between screens), and how the information or data collected is presented. It's up to you to pick the right method for your app idea. The only way that you can do this successfully is by knowing your audience first and foremost, but also by knowing what's possible in terms of navigation.

Those of you with a background in web design are likely to have delved into user experience design before, but in terms of an iPhone, there are many other factors that need to be considered. Apple are UX masters and are keen to stress its importance throughout their human interface guidelines. The iPhone OS is almost lyrical. Subtle animations and real world movements combine to create something so easy and natural to use, you sometimes forget it's not an extension of your hand.

iOS 7 has in-depth UX at its core. From the parallax effect on the home screen, which makes the icons appear to float, to the new animations, which drill down into an app to give the user the impression of delving deeper into the required information, the whole feel of iOS 7 is that your phone is alive. While the phone itself maybe flat, the design makes people feel that its depth is infinite. Whether you're trying to make an app feel wholly personal, or to give the user the impression that they have a whole world of information in the palm of their hand, tricks and styles you employ in this section can make that a reality.

Navigation Elements and Styles

iPhones are small. Though the iPhone 5 has given us a few more pixels of space, a well-designed app is a testament to your design skills due the incredibly small amount of real estate on offer.

But how do you manage to fit your idea into such a small space while still keeping the functionality clear? Learning the navigation and content organization views on offer is one of the cornerstones to good UX design. Each of your screens has a job to do, and it's your job to choose the right navigation style to help it to action that task properly.

It's unlikely that your app will be presented in a single view. Few and far between, apps such as the simple and useful Torch application manage to get to the point, and fast, but they don't do much else. There's not even a settings page. Many apps will have one main screen in which the action takes place and a utility view for settings or content libraries. Apple's Notes app has a utility view, one screen for the function and event, and another screen which houses all of your data. But whether your app needs one, two, or several screens, each app uses a series of view controllers to allow users to access information as quickly and easily as possible. Most apps will use a number of navigational methods to create a more rounded experience for the user.

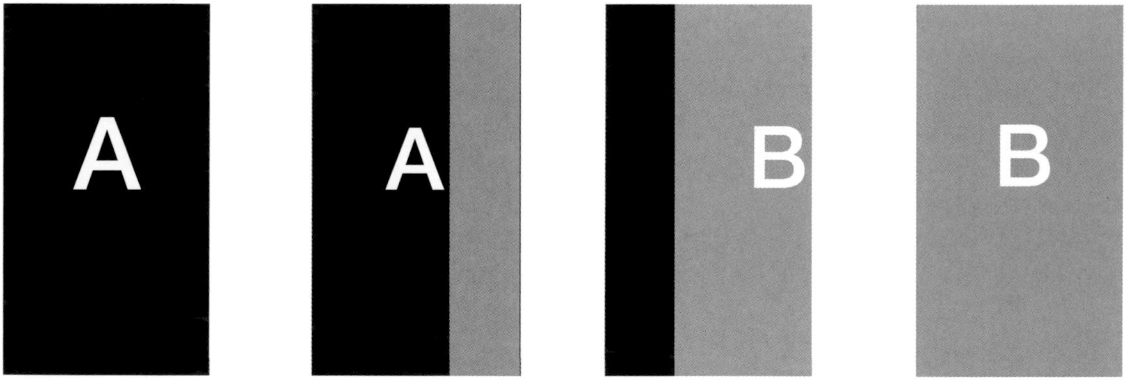

A

A

B

B

One basic premise to understand before we begin is that you either "PUSH" or "PRESENT" a screen.

When a view is "pushed," the current content slides from the right to the left. It appears to the user that you are progressing further into an app, and that the data is stacking up on top of the app's home screen. iOS 7 saw the introduction of Zoom to the batch of standard animations, which makes it appear that you burrow deeper into an app level by level. The calendar is a great example of this. You're then able to go back and forth between the data, or in iOS 7, to zoom in and out. When you tap on a text in your inbox and the message appears, the view containing the message has been pushed. A pushed view will never prevent interaction with the rest of the app.

To "present" a screen is to have it appear on top of the current view. Usually, this transition is animated and the new screen appears from the top or bottom of the screen. Other Apple animations for presented views include dissolve, or flip. The keyboards you use to input data are presented. The designer must factor in the size of the keyboard, understanding the impact it will have on the rest of the screen, taking into consideration that you'll be able to see less of the screen in the iPhone 4.

Though it's occasionally necessary to present a screen, a user will have to manually close that screen either by pressing done or using a similar button. Though this might not sound like a problem, having to close many screens in a session disrupts the flow of an app. They stop you from interacting with the main body of the app until the screen is closed. Present screens only when necessary to maintain a fluid feel.

It's important to learn the difference between the two presentation styles so that you know when best to implement each type of delivery and factor that into your flow.

iOS View Controllers or Navigation

We're now going to go through the different navigational devices that you have at your disposal. Many of these are included in Apple's catalog of standard elements, which we talked about at the start of the book. Remember that each of these elements can be customized to create something more personal.

Note down navigational styles that you think would be suited to particular functions or screens of your app. If you find nothing to suit, we'll explore what it means to create custom navigation at the end of this section.

NAVIGATION BAR CONTROLLER

This navigational method employs the "push" method. Utilized by most apps, it's the bar that appears below the status bar. This bar gives the user a little navigational context by showing them where they've been via the buttons present on the bar. Often, one of the buttons will display the word "back" as you progress in the app, but it can also display icons or alternative text which alerts the user to what will happen when you press it. Many app designers place access to settings in the navigation bar, or the ability to "add" via the small plus sign which appears in the corner. Here users can add contacts, or start a new note, for example. This is a convention you should bear in mind, as users are used to finding this information here. This recalls the familiarity aspect we've discussed. By applying this convention to your design, users are more likely to understand how to use it instantly.

There are many standard Apple buttons available for you to use in the navigation bar, and to customize if you so wish.

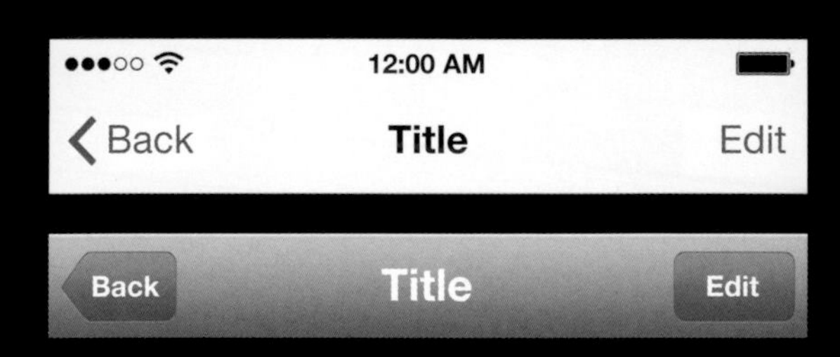

TAB BAR CONTROLLER

Found in apps such as Amazon and Twitter, the tab bar controller is a bar that runs across the bottom of a screen. This is great for quickly flitting between different screens of information, as it presents a self-contained mode of access to certain functions. Apps such as the Clock contain lots of mini-apps under one main umbrella through the tab bar: world clock, stop watch, alarm, and timer. You should not include more than five tabs in your tab bar. If you need to add more, you'll need to have four tabs, and then leave the fifth for a "more" icon. This will then take you to a view with the extra functions. The maximum pixel width for each tab is 64 pixels. The trick to using this navigation method well is in naming or creating icons that best represent the data or function that lies behind it. The most heavily used tabs should be placed to the left, as most people will look here first.

SEARCH AND SCOPE BARS

Got lots of data? If the user will need to search for something, this will affect your navigation. The scope bar allows you to define elements of the search. Mail uses a search and a scope bar, with the scope bar being used to define which aspect you're searching: To, From, Subject, or All. Some people get round the space required for the search bar by hiding it under the navigation bar. It only becomes visible when the user scrolls the page down slightly.

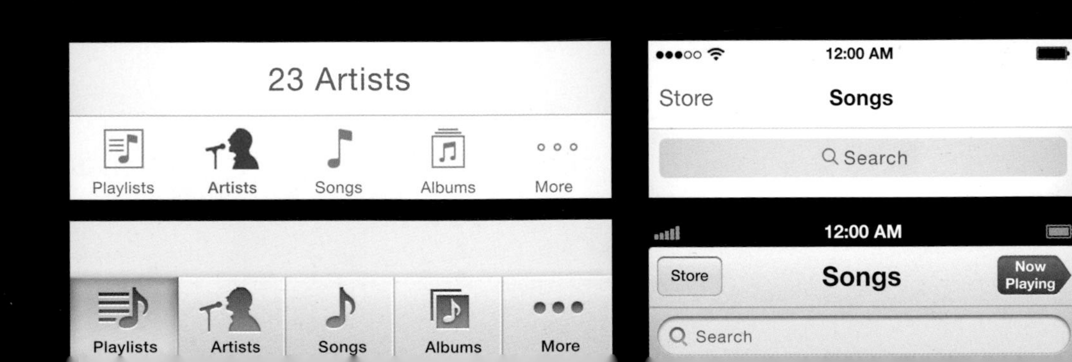

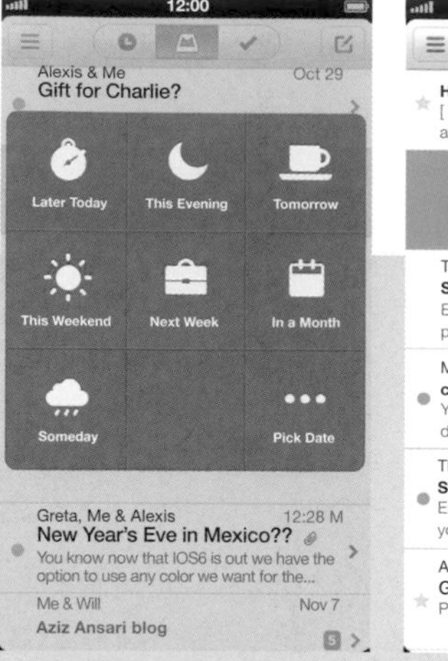

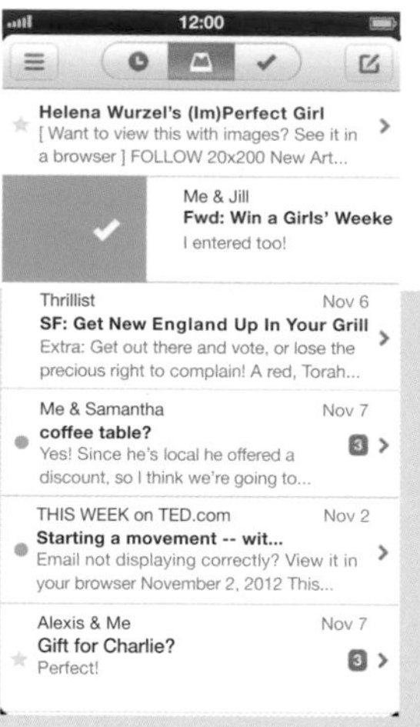

HAMBURGER CONTROLLER

The hamburger controller has risen to prominence in apps such as Facebook. It's actually a custom navigation device, only available as a standard Apple navigational method on the iPad, but many people are using it in their iPhone apps already. As the name suggests, you can interact or at least see, two screen of the app at once. While this might sound like a utility view, it's not, as all the action happens on one screen. A secondary screen is presented from the side. In terms of the Facebook app, a table view menu with the user's pages, groups, and more appear from the left and sits over the main screen until the user slides it back. Swipe to access menus are a similar variation of this. They create a more user-friendly navigational approach, as the extraneous information that the user needs frequently, but not all the time, is housed neatly on the main page. The hamburger controller makes for a speedier app. Keep in mind that this is a custom navigational view and you should talk to your developer before you decide to build it into your design.

CUSTOM

Of course, there are times when traditional navigational systems just won't do, and in turn developers create custom navigational methods. At one time, the Facebook app was such, but the experience fluidity it provides has meant that its navigation method has been replicated by app designers all over the world. Creating custom navigation is a risk as you're steering away from conventions that people know and understand, but some app developers have made their name through successful custom navigation.

The Mailbox app managed to do the seemingly impossible, by reinventing the email app. The user can do almost everything (except read the message!) from one screen. To archive, you swipe a message to the right. To delete, swipe longer in that direction. Create reminders and lists by swiping similarly to the left. The navigation is intuitive and easy to learn, and it makes clearing an inbox infinitely quicker.

Remember that all of the standard navigational elements can be customized by you or your developer. Before you set about planning a new navigational system, make sure that you can't achieve the same effect by modifying some of the tried and tested methods and elements currently out there. Talk it through with your programmer. It's better than putting a load of work in only to be told it can't be done, or worse, that there was a better option out there.
The same goes for animations. If you want to present a modal view with the screen hatching from an egg which appears from the bottom of the screen, you might want to talk to your developer about that first!

Though it might seem alien at first, looking through the apps that you have on your own phone and observing the different ways the designers have moved between screens can teach you a lot. Look closely at the apps you find easiest to use and use most regularly. You'll be so tuned in to the way that you navigate your way through these that you'll probably be more familiar with navigational methods than you think.

Other things to consider:

STATUS BAR

Though you might like the extra few pixels, your users like the status bar. One of the major changes in iOS 7 is that all apps are now full screen as the status bar is transparent. You'll need to make sure you still leave space for it, and extend the background up beyond it.

12:00 AM

12:00 AM

Label Title +

Search

Best practices become obvious when you look a little deeper at the good apps already out there.

Temporary Views

You won't be able to change the visual appearance of all of these, but it's important to plan for them. If the temporary view covers information vital to the task you're doing, you're creating a poor experience for the user. Make sure you factor in temporary views when planning the flow of your app.

MODAL VIEWS

Modal views appear on top of the app screen, temporarily preventing action or further functionality with the rest of the app. Modal views are presented. This could take the form of a pop up (Apple aren't fans of these; use only when absolutely necessary), a keyboard, or a text box. The buttons which allow you to email a picture stored in your photo gallery are modal controls, as they appear above the image, and are presented from the bottom of the screen.

Modal views are great for extracting data from the user or for adding extra functionality (screens such as "edit," "add contact," or "change settings"), but should be used sparingly as they cover up the main action. Modal views are particularly good for things like notifications, and when you want to focus the user's attention on a piece of important information, as they are effectively removing the user from the flow of your app.

ACTION SHEETS

These are used to display a set of choices to the user, i.e. "save" or "delete." Though you can change the text and colors within them, you can't customize them any further than this.

Set Lock Screen
Set Home Screen
Set Both
Cancel

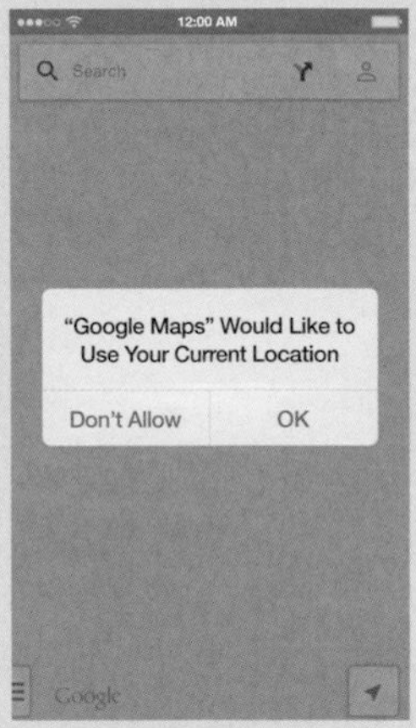

12:00 AM
Search
"Google Maps" Would Like to Use Your Current Location
Don't Allow
OK

ALERTS

Alerts are questions or settings that the user needs to action, such as "allow Facebook to determine your current location" or "Do you want to save the image before closing?" These always look the same, regardless of your app design, and are more of a user experience consideration than a design worry.

Controls

This set of UI elements allow the user to interact with or extract information from your app, including buttons, page controls, sliders, and switches.

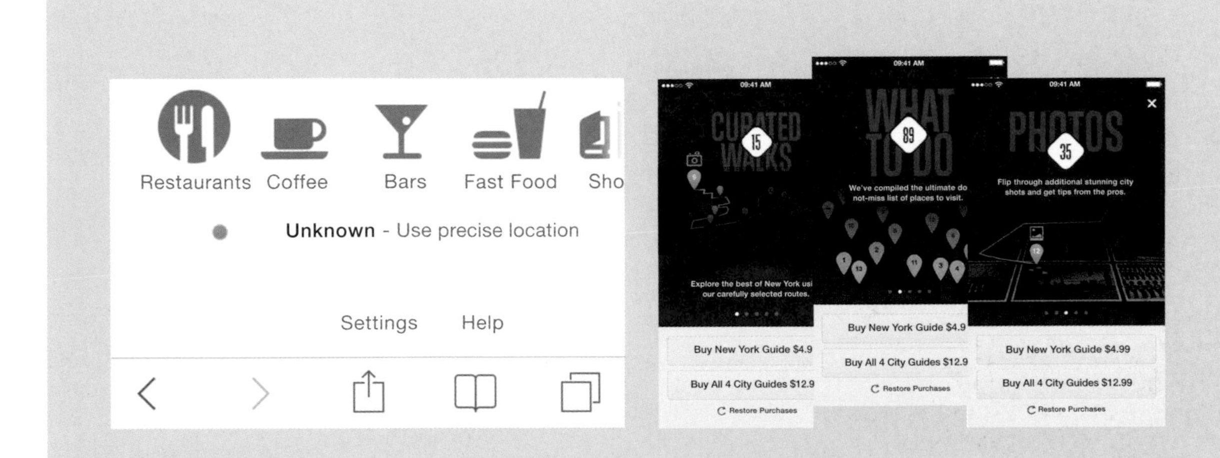

TOOLBARS

Toolbars provide a container for buttons that allow the user to perform certain tasks on the page, such as "delete" or "reply." Unlike action sheets, which appear over the page, toolbars are housed on the page, often at the bottom in the same place as the tab bar. An example of a toolbar is found in Apple's Photos. When you look at a single image, there is a toolbar at the bottom which allows you to "share," "start slideshow," or "delete" the image. Pressing the icons housed in the toolbar won't take you anywhere, it will allow you to perform an action or to enter some kind of modal view which will let you take further action there. Apps such as Safari have their back and forward buttons housed in it. Social media and share buttons are often found in the tab bar.

PAGE CONTROL

This method of control is found in apps such as the weather app, Swacket, and the new iOS 7 Compass app, which also includes an inclinometer. It's very useful for displaying multiple pages of content across one level. You'll see dots at the bottom of the page, which represent the number of screens. Used with the scroll function, the user can then move between the information by moving a screen to the side by swiping and flicking. It's also a great device to use for first-load walkthroughs; as demonstrated in the National Geographic City Guides app (above), users can scroll through the instructions or feature list. As soon as you see the dots, you know that there's content there to look through.

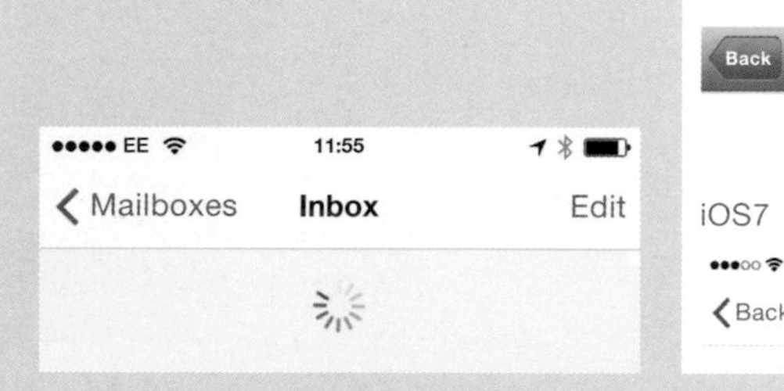

DATE PICKER

This is used in Apple's Calendar and Alarm app. Instead of leaving the user to input the date or relevant data themselves, you provide them with choices that they then select from to action the function. In iOS 7, in apps such as Calendar, Apple have begun embedding the picker into the content, having it appear when required, so that people can remain on the same page to create a more fluid experience.

REFRESH CONTROL

We've mentioned this one before when we explored interaction design. Found in Apple's Mail app and Twitter applications, users can pull down on the screen to action a user-initiated refresh. While commonly used in many apps, it should only be used in apps that require frequent refreshes. Also bear in mind that the user will be returned to the top of the screen when the data is refreshed. This isn't always what the user wants, especially if they've trawled their way down a long feed.

BUTTONS

It would be slightly condescending to explain this method of control. It's a button.

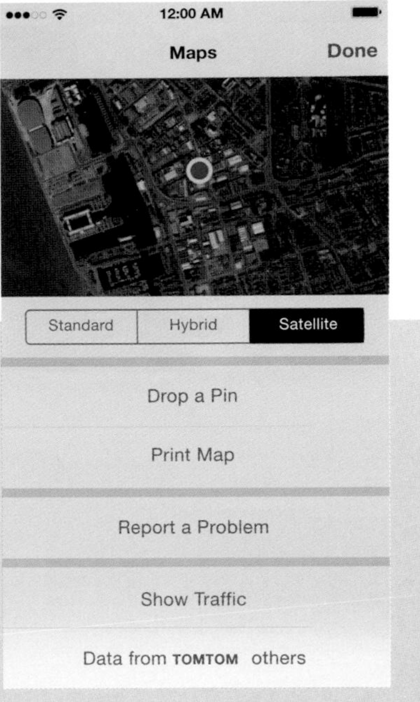

SEGMENTED CONTROL

You can find this one in Apple's Phone app. The control mechanism at the top of the screen that allows you to organize calls into "Missed" or "All" is an example of a segmented control. They're useful for showing or hiding data without changing the current view, making your app more streamlined.

SLIDERS

These are used to make in-app adjustments, an example of which being when you configure the brightness of your screen. Sliders are commonly used in the accessibility options, to let the user define the size of the text.

SWITCH

Another obvious one, use this to change between two choices, i.e. "on" or "off." It's found frequently in settings menus.

Transitions and Animations

We've mentioned a few animations and transitions already in this chapter, but here we'll lay them out in more detail. Transitions are extremely important for the iPhone. Views rarely change straight from one screen to another. Instead they use methods that add to the user experience, such as turning a page in the same way as you do in the real world. Only use a transition when it makes sense contextually.

Common transitions:

FLIP

This transition "flips" the screen to the next one as if turning over a card. It's not the fasted and smoothest transition, but it certainly has its uses.

ZOOM

Used on the iOS 7 home screen, it's this animation that zooms into the application, appearing to take the user deeper with every touch. You can also zoom out when the user needs to go back.

DISSOLVE

This transition allows you to fade a screen in or out. Kind of like when your character dies in an '80s video game.

PAGE TURN

Used in iBooks and digital magazines, the page turn is self-explanatory. Try to only use this in a situation where you would in fact turn a page. It might seem a bit weird otherwise.

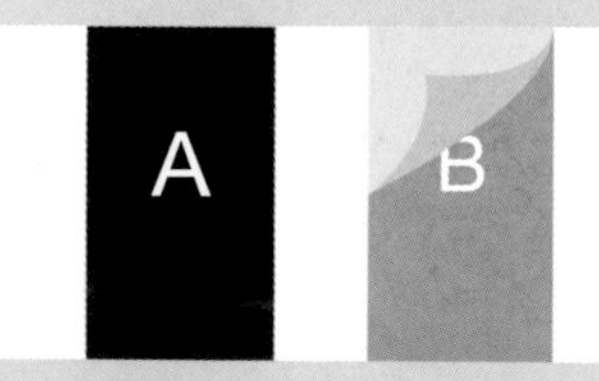

PUSH VIEW

A push view animation is the most common as it's the default animation for presenting a new screen. The animation slides a new screen from right to left to add it on to a stack, like a pack of cards. Users can move back to the previous screen by swiping from left of the screen to the right, or by using the back button automatically placed in the navigation bar top left.

iOS 7 gave developers a few new APIs to play with which can enhance the look of an animation by making subtle changes to an overlapped view. This can take the form of a blur or gradient change. As users can now swipe to return to the previous screen as well as using a button, these subtle changes can be used to a designer's advantage. For instance, you could ask your developer to completely blur the first screen as the second screen is revealed over the top. Apple use the effect subtly in the navigation bar and table views as you move between screens.

Sometimes you'll want to animate certain functions to draw attention to them, either to educate (many developers employ animation on a user's first-time run) or to highlight that there's something new. The important thing is to make sure that your animations don't detract from the real reason that your user is there. Talk it through with your developer to see if they think it adds value.

Use methods which add to the user experience and only use a transition when it makes sense contextually.

Content Views

There are many different types of content views that you can use to organize the information that will appear on a screen. From collection views to map screens, looking over this brief list may give you extra functionality ideas that will supplement your app.

TABLE VIEW CONTROLLERS

These are used by most serious apps to display data. Though the table view controller is rarely seen without either a tab bar or navigation bar to supplement the action, it is a navigational method in its own right. Mail, Messages, Notes, and Voice Memos all use table views to display the user's data, and clicking on a cell in the table takes you through to that content. Data in table views is generally pushed.

12:00 AM
Settings Photos & Camera
PHOTOS
My Photo Stream
Automatically upload new photos and send them to all of your iCloud devices when connected to Wi-Fi.
Photo Sharing
Create photo streams to share with other people, or subscribe to other people's shared photo streams.
SLIDESHOW
Play Each Side For 3 Seconds
Repeat
Shuffle

12:00 AM
Groups All Contacts +
Search
#
John Appleseed
Kate Bell
Anna Haro
Daniel Higgins Jr.
David Taylor
Hank M. Zakroff

ACTIVITY VIEW

Generally a modal function, this view appears above the current view to show the extra services available to the user, such as social media sharing, email, and printing. Apple's own functions such as AirDrop are also displayed to the user here.

It's not terribly important that you learn the exact names of all these displays if you're not programming the app yourself; the important thing is to know they are there. Just reading through them should have given you some ideas of how to present your app. Note any that fit with what you're trying to achieve; they'll come in useful when sketching and wireframing.

PAGE VIEW

Found in apps such as iBooks, page view allows you to move between defined pages of content by either turning pages or scrolling. It makes sense in the context of a book or similar, but not much else.

COLLECTION VIEW

This presents an organizable collection of data for the user. Whereas table view is a list, collection view is a grid of images or icons which can be used to navigate. An example of collection views in use is your photograph albums. They fill up the screen with the user's data so it makes the user feel like the app belongs to them. Try to show one type of data at a time to avoid overwhelming your user.

MAP VIEW

Map view displays a map which works with the majority of the phone's inbuilt map functionality.

IMAGE VIEW

Use this to display an image, i.e. display the photographs you've taken with the Photo Booth app.

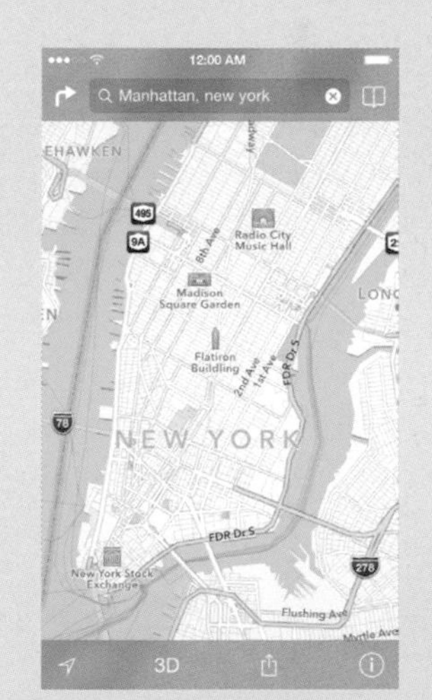

Settings

Before you start planning out your app, figure out whether your app will need a settings section, or if you'll need to allow users to modify aspects in-app. Will your app allow the user to change various elements of your application to suit their needs? Settings can be a great way to personalize an app experience, but in some cases they can actually be a hindrance.

Settings can either be found on the page (or inline), accessible via buttons or a modal view, in the settings section (outside of the app), or on a separate page in the application, usually accessible by a navigation bar or small cog icon found on screen.

If a setting will need to be changed frequently, then it's not a good idea to put it into the Apple Settings section. It would disrupt the flow of the app and irritate the user. In this case, the setting should be changed in-app, either on its own page or via a button or modal screen.

If the user will rarely need to change a setting, then you might be able to put it in the Apple Settings page. Talk it through with your developer. The aim is not to overcrowd your page with decisions. To make something intuitive, there has to be a clear path for the user to follow. With the exception of the actual settings page, giving eight different settings options on a normal screen isn't the definition of good user experience design.

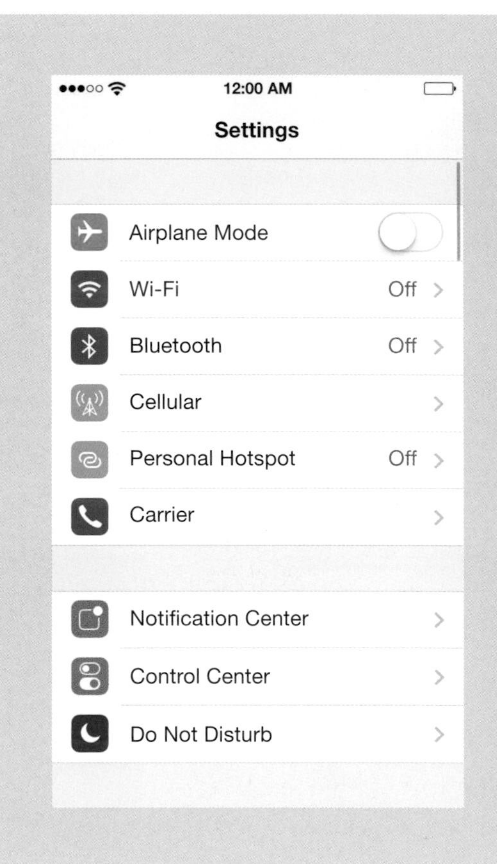

Use Case Scenarios

Before we move onto sketching out your app, we're going to look at an exercise which will help you design the perfect UX. We've touched briefly on case scenarios before. When planning your app, we suggested that you think about the many circumstances in which people will use it.

In terms of UX design, it's important to recognize that people have different needs in different circumstances, and for the best experience for all users, these scenarios need to be addressed. The outcome of your user case scenarios will almost certainly indicate the type of navigational style you should use and help to define other aspects of your UI.

To do this, you need to put yourself in the shoes of the potential user. If your app is for everyone, you'll need to put yourself in the very small shoes of a child as well. Think of the circumstances you envisage people using your app. It might help to brainstorm the different places, things, or people you'd associate with your app's subject matter.

A simple example is that of Apple's own Calculator app.

Case scenario one:
Vicky wants to split a bill at a restaurant. She'll only need to use the "add" and "divide" functions. Everyone's had a bit to drink and extra functionality would be a hindrance at this point.

Case scenario two:
Paul has forgotten his scientific calculator but needs to work out some complicated equations. A basic calculator does not have the functionality he requires at this point, so he needs to use his phone.

Apple's Calculator, despite being a basic app, has UX at its core. Case scenario one is covered by the portrait view of the app, which has basic functionality, while case two is solved by the landscape view, which boasts scientific calculator functionality. The way Apple has designed it means that users who don't need the extra functions need not be bothered by the more cluttered interface of the scientific calculator. The portrait view has larger buttons than the landscape mode. As it's unlikely that people will be solving scientific equation on the move, smaller buttons won't pose a problem for people using the calculator in landscape mode.

Try and put yourself in the place of your users by thinking up the various situations that your app might be used in. It's only ever by fully understanding your audience that you can solve the problems that they might have.

Case scenarios shouldn't just look at the obvious. You should also think about situations such as closing an app unexpectedly after spending a long time on a function. What happens then? If your app requires internet connectivity, what happens if you lose signal?

The case scenarios will not only improve on your idea, they will alert you to screens you might not realize you needed. How about this one: has your user misspelled their password? You'll need an error screen for that. You might need another which allows him or her to enter their email address to choose a new one.

Case scenario: Harry is about to make a purchase and needs to enter his password, but he realizes after one attempt that he has forgotten it.

Say your app is like the Mailbox app. A case scenario (right) that the developers solved was the following: when all of the emails in the Mailbox app have gone, the user could have been left looking at an empty screen. Instead, the developers chose to take the user to a screen which says "You're all done," which contributes to a better UX. The user feels they have achieved something, and to boot, it's better design.

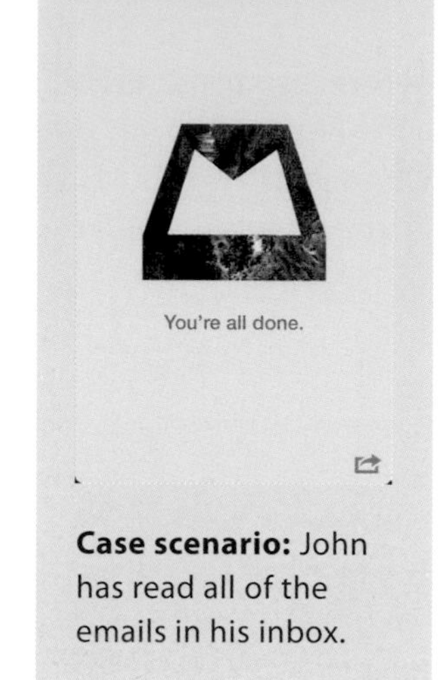

Case scenario: John has read all of the emails in his inbox.

You have your app idea. This exercise gives you the opportunity to logically work your way through the how, when, and where, but also the "what happens if?" Make sure you write down the outcomes, modifying your list of functions and features as you do it. It's also worth noting down the various screens you'll need. Though you'll undoubtedly add more during prototyping, it will act as a guide.

Paper Prototyping and Defining the Flow of the App

This section will take you from sketching through to a paper prototype. We've added some resources for you to download at www.iphoneappdesignmanual.com.

Though you might be keen at this point to set up your iPhone template and get straight to the digital design, pen and paper are in fact your best friends. Whether you draw boxes or use a paper template to sketch out your ideas is irrelevant, what is important is that paper prototyping gives you the opportunity to fully understand any problems with your app idea. You don't need to be able to draw, all you'll really be drawing is basic shapes, so don't let the fact you're not Picasso put you off.

It might seem quicker to start laying out everything in software, but the simplicity of paper prototyping allows designers to think more clearly due to the lack of detail. Ideas that might have remained dormant in the face of an almost finished design spring to life when looking at the barebones of an idea.

When we write notes in word-processing software, it's unusual to delete a piece after we've worked on it for a while, even if it's not living up to expectation. Instead, we'll amend and delete, add new passages until it cuts the mustard. In a similar fashion, if we've spent time creating a digital prototype or highly detailed sketch, we're likely to try and fix it even if it's not working. We amend what's already there instead of starting afresh like we probably should. The beauty of paper prototyping is that if an idea isn't working out, you've spent minimal time sketching it and you'll be inclined to put it to one side and start again.

Sketching also has other advantages. If you create a beautifully detailed Photoshop image of your app, people who you show it to for a second opinion will be less inclined to criticize

as it looks finished. A sketch is open to suggestion, and will prompt more feedback. Be prepared to sketch a lot of variations.

Each screen is called a "view." The main screen is the "root view" from which all other views will stem. You could draw the root view first and use arrows to show which view comes from each active element (toolbars and buttons) of your root view. This makes it easier to understand your flow later. Some actionable areas will only take you to one screen. Some will take you on to many. Having a big enough sketchpad to accommodate this journey can make things easier. That being said, you can always join pieces of paper together. Just do it in a way you find logical, as your finished sketch will form the diagram from which you'll create your wireframe. Keep in mind that every view needs to come from somewhere and will appear from some action by the user. If you don't build the button, nobody's getting to that page.

Cross off functions you'll be using from your list to make sure that you get them all. If there will be text you can denote this with a few squiggly lines in your sketches. There are a few considerations before you start. Some apps will run differently the first time they start up to get required information and settings in order. If your app requires you to input personal information to allow you to use it, you should be aiming to get this information as soon as possible so that people can get started with the main functionality. Good apps only require you to input personal information or allow settings (such as push notifications or to use your current location) once. This information can be changed by the user later in the Settings menu. The first-time run sequence will eventually land back on your app home screen, so you start your normal run sequence from the home view. Note down which of the screens will appear on the first run only.

As we said earlier in the chapter, each screen needs to serve a purpose. If your app is based on user data or preferences, you don't want your user to land on an empty screen on the first run. Factor in asking questions, which will populate the main screen, before they first get to it. This way the app is functional right the way through the first run and it's a more fluid approach.

Many designers use sticky notes so that they can swap and add screens to an overall layout. At Apposing, we have stacks of paper templates ready and waiting to fill with our ideas. This way, however simple our sketches may be, they're to scale and we have an instant understanding of whether the app is looking too crowded. You can download our templates from our website.

PAPER PROTOTYPING TIPS

Now that you've got to grips with interaction design, and have an understanding of navigational methods, it's time to pick up a pencil and watch the app take shape. These sketches will be a great point of reference when talking to your developer.

- Take your list of final functions and using the home view as a start point, map out with pen or pencil how the user will move from one screen to the next. You don't need to focus too much on the navigational methods at first, rather which screen follows which. This way you can get your house in order. What you're aiming to do here is not to create a replica of how the final thing will look, rather the bare bones of how people will achieve the objective.

- Use arrows to show which screen follows another and make notes in case you forget mid-sketch ideas. Try to note down gestures with arrows and lines.

- Remember that you don't need much detail in your sketches, nor does the UX have to be fully formed at this point. Even if you haven't fully figured out an action, sketch down what you do have. The ambiguity of your early sketches can be a great breeding ground for your best ideas. Try to work through your app idea logically. Sometimes a user will be able to go in two directions, so you need to stay focused on each of the app's functions when mapping out the options.

- When you're done, it's wise to check the flow. If you don't fancy doing it manually there's a great app tool called POP. It allows you to take pictures of your sketches and link them together via "link spots" which you define on the image. You can then go through your sketches to make sure that the flow of your app is correct, and you'll be left with a very early stage prototype (and a storyboard, which we're just getting on to . . .).

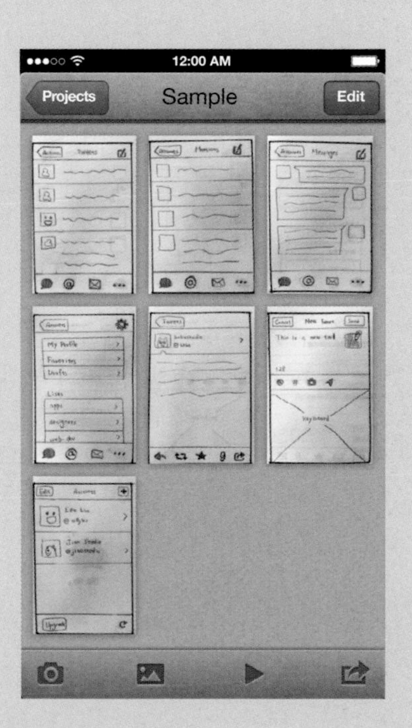

Wireframing and Storyboarding

Finished sketches in hand, you can now move on to something that's a little more high fidelity. It's time to create a wireframe for your views, and a storyboard for your app.

Wireframing

Wireframing is similar to paper prototyping: it's visually laying out your screen, albeit basically. There are an abundance of great wireframing tools out there which will allow you to create each of your views to exact specifications. Though paper prototyping is great for sorting out the flow of your app, exact dimensions aren't playing heavy on your mind, and rightly so. But screen size is limited, and by mocking up a very basic wireframe, factoring in the exact specifications of elements such as the tab and navigation bar, you'll have a better idea of whether the screen is overcrowded. You can also add text to buttons here, giving you more visual context. Tools we use here include the offline software OmniGraffle and Moqups for online wireframing. There's also Balsamiq, one of the better known and more widely used choices. Balsamiq requires a little working knowledge of its processes to be able to use it effectively, so you'll need a little practice. If you haven't got time to learn the basics, you might want to opt for a less advanced wireframing tool. (Of course you can use Photoshop, but these tools offer benefits. If you are using Photoshop, check out DevRocket, a useful plugin to make your job easier.)

With dedicated wireframing tools you'll be provided with to-scale templates of each of the iPhone variations, and some also have icon-sketching features, which show how your icon will look in all size variations (although worry about icons in chapter eight!). Many also provide to-scale Apple navigational elements. Though you can easily find out the sizes for each element (status bar, navigation bar, etc.) these tools will do that for you. You'll need to give special consideration to how content built for the iPhone 5 will resize for the iPhone 4. It's not always just a matter of shrinking to fit.

Storyboarding

Storyboarding is a visual representation of how the screens in your app relate to each other. It's used by folk in TV and film production to plan out each individual scene in their feature. That's exactly what we'll be doing here. A storyboard is a series of wireframes or sketches in order with arrows and notes to show navigation between each view; it's essentially a user interface flow diagram. You'll be left with a simple image of how each screen links to the other.

Xcode, the mac software you or your developer will use to program the app, includes a piece of software called Storyboard. Not only can you mock up the flow of an app in Storyboard, you'll also be left with a simple interactive working app that you can use on a simulator or your phone. While you will need Xcode to use it, you don't need to write a single line of code.

This part of the app development process might be the most stressful, especially if you've not built an app before. It takes time to understand the nuances of user experience design and to know which navigational view will work best. When this stage is done, you'll be let loose on the final design of your app, so take the time to get your storyboard right or you'll find you pay the price later.

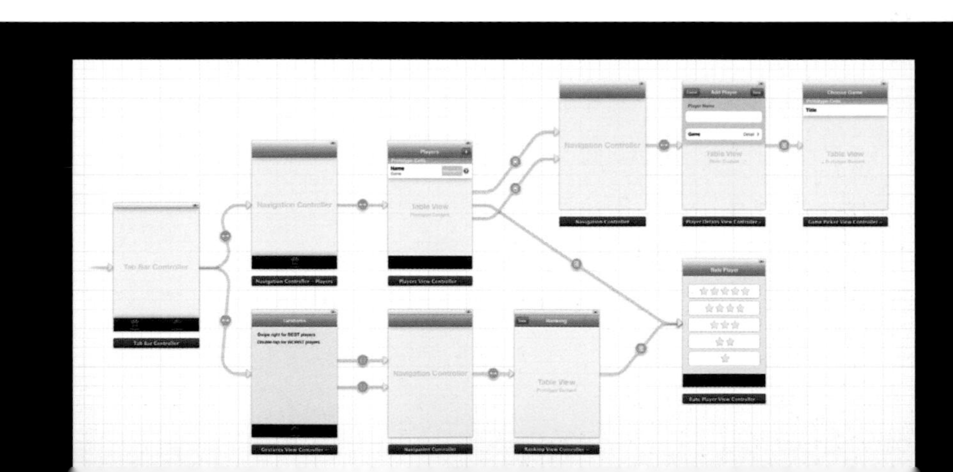

CHAPTER SEVEN

INTERFACE DESIGN

If you've got any previous design experience, then it's this part of the journey that might appeal to you the most. Though there will be many new considerations outlined in this chapter for you to integrate into your current design knowledge, it's during this stage that your idea gets a visual representation. Your concept needs room to breathe; nothing confuses a user more than an overcrowded screen. Your design also needs to make using your app intuitive; there will be no instruction manual. Good interface design takes all of this into consideration, making things look esthetically pleasing at the same time. It's no mean feat, but this chapter outlines our top tips for success.

What is Interface Design?

Put simply, interface design is the app's face, the part of your product that users will interact with and explore. Not to be confused with interaction design, which we explored in chapter five, interface design presents the controls with which a user can explore an app. This part of the process is purely visual.

Interface, interaction, and user experience design are closely linked. One can't work efficiently without the others. Now that you have your interaction and user experience design planned, it's time to turn your work into an intuitive interface that will be a pleasure to look at. Users will look to your interface design to see what is most important, and what they can and can't do. It's your job to make these choices obvious to them. Read Apple's Human Interface Guidelines (see references). They outline in great detail the nuances of the process and, more importantly, what's expected of you. Unlike web design, app design is not as open-ended. Though conforming to expectations might seem alien to the achiever in you, when it comes to app design, the codes and conventions are there for a reason, most of which centers around creating a fluid experience for the end user. The industry has rapidly expanded because apps offer convenient methods of making life easier and more enjoyable. Sticking to these guidelines ensures that your app doesn't require a new mindset, or set of instructions. Consistency allows people to take what they've learned and apply it to your app without hesitation, strengthening affinity between app and user instantly.

1 Elevate important content: Make it easy for your users and place the app's primary function in plain sight. The upper half of the screen is often a good place to focus attention. Make sure that things are easy to read from left to right too; it's the way that the Western word naturally reads. However, if your target audience naturally reads from right to left, factor this into your design. Don't make your users search or scroll for the thing that defines your app.

2 Keep your app's primary purpose in mind: If it's an entertainment app or game, you have a duty to make it beautiful and originality is applauded. Likewise, if your app is task-driven or information-heavy, you need to make it easy for the user to achieve their goal, so function must always be at the forefront. Both types of app require the look to match the function; give your user the experience you intended by ensuring this is the case.

3 Leave the invention to your concept: While reinventing the UI wheel may seem like a great way to make your mark on the app world, a great idea could be marred by an overambitious interface that doesn't conform to any of Apple's guidelines or make sense to the end user. They didn't write the guidelines to keep developers in check; they created them because they know a lot about what's best for their customers. People like knowing what to do instantly. Make it easy for them.

4 Minimalism rules: If it absolutely doesn't need to be there, then get rid of it. Excess design is messy and hinders the flow of your app. iOS 7 placed a greater emphasis on white space than previous iPhone iterations. Keep it clean and considered to make your design shine.

5 Instinctive controls are key: Your app's navigation should be at the focus of your interface design. There's no point making a UI that's stunningly beautiful, but leaves the user unsure of how to get from A to B.

6 Offer encouragement: There's no harm in letting people know that they're getting it right. File downloaded successfully? Let the user know. All tasks on a to-do list completed? Say congratulations. We're all suckers for being told we're right. Also think about calls to action: if a user needs to enter information to use a particular function, ask them for it so that they're making the most of an app.

7 Context is key: This one seems pretty obvious but it's amazing how many app designers forgo this rule when putting an app's UI together. If you've followed our advice, you should have a clear idea of who is using your app and where. Keep this context in mind and design specifically for your target audience and the way in which they'll be using it.

8 Don't innovate for innovation's sake: Some things don't need too much embellishment. While inventing a wacky way for your user to delete a message may seem fun, in the long-term, it's unlikely that they will thank you for it. Tried and tested methods for mundane aspects of your app are fine, and won't detract from your app's core purpose.

9 Keep it consistent: Don't use a different font or color palette on each page, unless it's a fundamental part of your app. If your app is for an established company, follow the brand guidelines. If not, it doesn't hurt to think of your app as a brand. By giving it a distinctive look and feel that's consistent on every page you're creating a streamlined experience for the user.

10 It's a balancing act: It's natural to assume that the biggest thing on a screen is the most important. Don't confuse your users by giving a subsidiary function more prominence on the screen.

Custom Interface Considerations

Whether you're customizing Apple elements or designing your interaction and navigation methods from scratch, there are some basic size requirements and considerations that apply to both methods. You'll also need to think about how different screen orientations will affect your final design.

Customizing Apple Elements

The majority of you will be customizing Apple UI elements, such as the tab bar, buttons, and navigation bar. This is pretty easy as long as you know the exact sizes of the elements that you will be creating a new façade for.

There's a quick reference table at the back of the book that outlines how big each of these elements are. There's also a size reference for icons that appear in the tab bar.

Custom Design

If you've decided to go down the route of a completely custom-designed UI, then the sizes are really up to you. Remember to make sure that the minimum "tappable" area is 44 x 44 pixels. You can use the table we've provided at the back for reference, looking at the sizes that Apple use to give you a guide as to how big most people expect elements to be.

Landscape or Portrait?

The question is not really a matter of which orientation you want to work in, it should be a question of the user requirements. By offering your user multiple orientations, you're allowing them to decide which way is most comfortable for them. As outlined in the case scenarios section (see page 107), some apps use the different orientations to provide more or less content, as seen in the Apple's Calculator app.

If you're going to provide alternative functions or layouts in a different orientation, as you've probably assumed, you'll need to design them separately. However, if you'll be providing the same content in a different orientation, as is possible in the Mail app, and your developer is using Auto Layout, the content will snap into the right place (providing it has the correct constraints), stretching Apple elements without pixelating them on the alternative orientation.

The Future of Screen Sizes

With the launch of the iPhone 4 back in 2009, the introduction of the retina screen had a big impact on UI images. Designers had to make graphics double the size required for the non-retina screen, as each screen now contained double the pixels. Designers who didn't do this found that their graphics were pixelated on the retina screen.

If you've decided to support iOS 6, or if you're also designing for iPad, you'll need to export each of your graphics at two different sizes to cater for the differing screen capabilities. You'll need one for the non-retina, standard screen, and one double the size for the retina screen. We focus on exporting and delivering your design assets to the developer in chapter ten. At Apposing, we find it easier to size down, i.e. we make our graphics at the larger, higher-resolution size, and reduce them later. If you're designing for iPhone only and for iOS 7 and above, this won't apply to you as all applicable devices feature a retina screen.

There had been no changes to screen dimensions until the launch of the iPhone 5. If this is anything to go by, screens are getting bigger. Though we can't foresee a time that an iPhone won't fit in your pocket or an iPad in your bag, an extra centimeter of space can make a major difference to your app. Screen size changes can play havoc with design, but help is at hand . . .

Auto Layout

With iOS 6, Apple launched Cocoa Auto Layout for iOS developers. Developers choosing to build with Auto Layout find that the problems caused by differing screen sizes are resolved quicker. Auto Layout allows the user to define constraints for the application that are able to transition between screen sizes and iOS platforms. For instance, you can define that elements line up in a straight line. Previously, a different screen size or iOS would mean that your design may become unaligned or look bizarre when loaded on a different model or OS, however Auto Layout ensures that it will pretty much look the same across devices due to the rules you've put in place. As screen sizes change and UI values adapt, using Auto Layout will allow your app to respond accordingly.

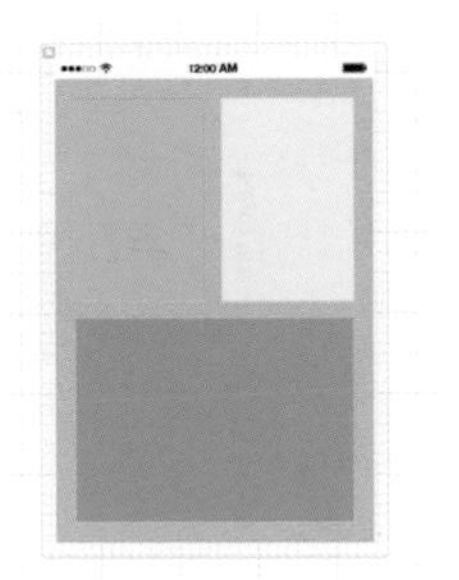

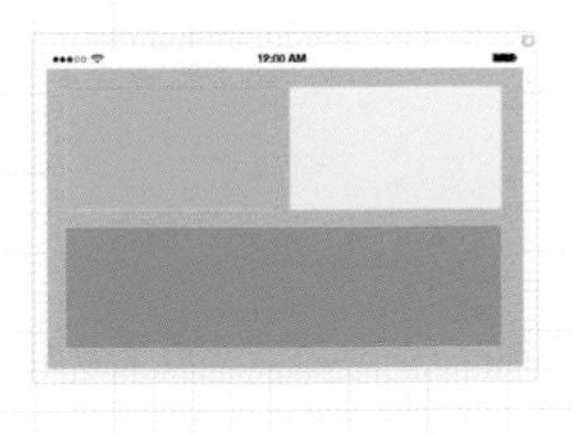

The Anatomy of an iPhone UI

Just to make sure that we're all on the same page, this diagram outlines the various design elements that your UI will feature.

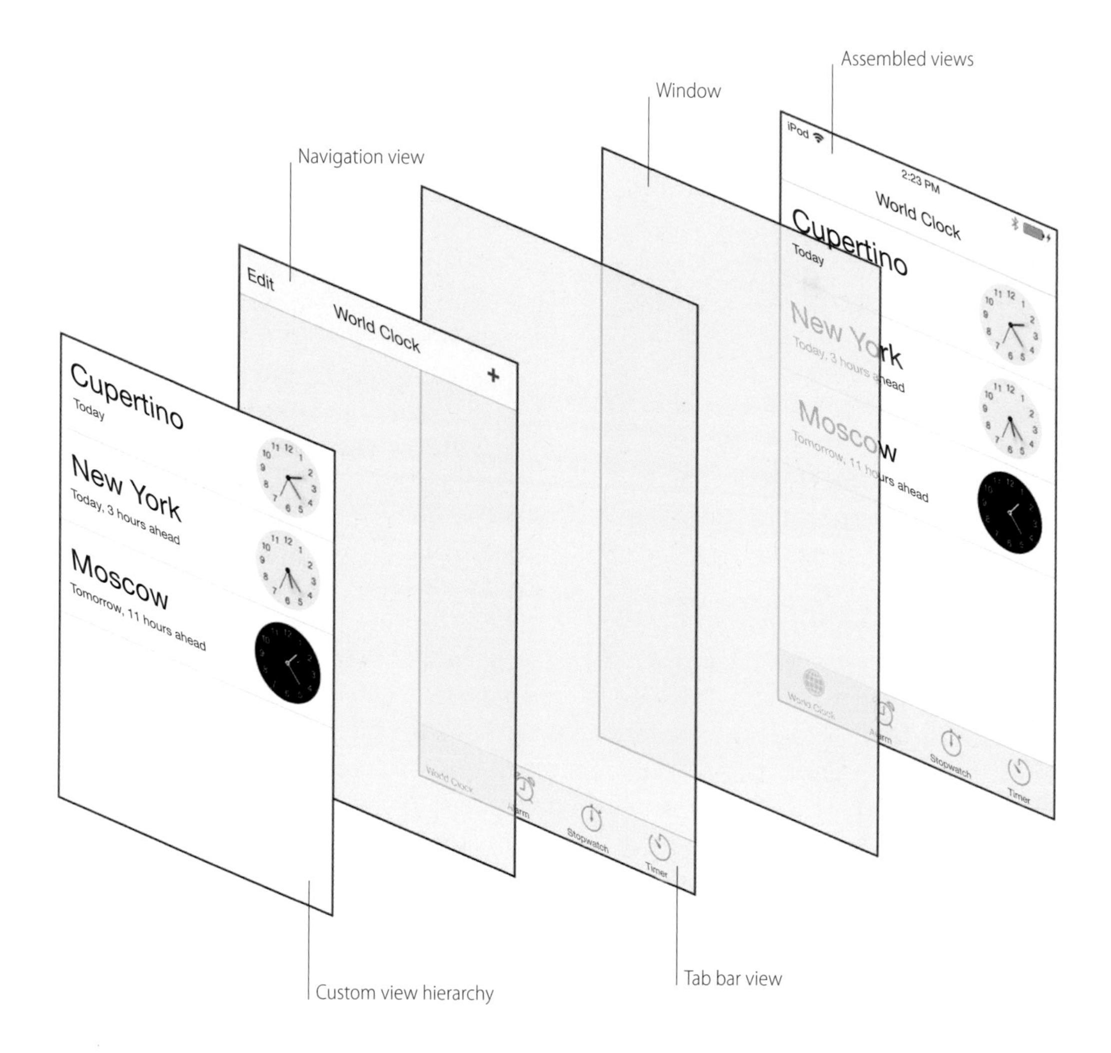

App Design is Changing . . .

The introduction of iOS 7 brought major changes to the App Store and the design process. If you've grown up with the iPhone, the device's penchant for the skeuomorph (a digital visual representation of a real world object) cannot have passed you by. The idea was that a visual metaphor would add to the intuitive nature of the app, as you'd already know how to turn "a page," or to press "a switch" to turn something on. But those days are behind us. The launch of iOS 7 tossed the skeuomorph in the metaphorical trash and replaced it with something a lot flatter.

7

Though it will always take some time for every iOS user owning an iPhone to transition over to the latest OS, ignoring any new design guidelines and rules would be extremely shortsighted. Your app will look dated from the offset. The good news is, those of us in possession of a developer license get access to the new OS months before the general public, meaning we've got time to future-proof our app against updates. It's unlikely that an update will take you by surprise.

Some of you will still want to support iOS 6, but in terms of design it's always best to look forward and we'd advise against it. Designing for backward compatibility is limiting (some of the latest iOS 7 design variants don't work with iOS 6). By having two versions of an app that differ in design, you're diluting your brand, when you really should be aiming for a unified experience across multiple devices. By using software such as Auto Layout (or getting your developer to), you can define values which will allow your app to look the same on multiple operating systems.

The visual difference between iOS 7 and iOS 6 was dramatic. Simplicity is key. We've stressed many times that you should remove any design or functionality you don't need, and this is good advice for dealing with iOS 7. The iOS 7 ethos is centered around giving you exactly the tools you need for a job and stripping back the rest. The simplicity that the skeuomorph was intended to bring in previous versions has been rethought. The new thinking states that you'll know how to use good design instinctively, without it needing to look like its real-world counterpart.

The fact remains that you have license to do whatever you want with your design. You can forgo the redesign completely if you so wish. While our team of designers might have had mixed feelings about an update initially, we fully embrace any changes and put forward its ethos to our clients. We don't want our icons to look out of place next to Apple icons on the home screen. iOS 7 is clean, modern, and, for now, offers designers a chance to shine by embracing the new "flat" trend. It's so open at the moment that anything is possible.

Don't view the death of the skeuomorph as something to grieve over, you're no longer tied to real-world objects. Let your design skills shine.

KEY THEMES OF APPLE'S LATEST LOOK:

1 Deference: The UI should guide the user as to how to use the app while never competing with the content. You need to be aiming for simple design that lets the function shine. Always respect the function.

2 Clarity: From icon to font, everything is crystal clear and legible at all times, with functionality at the forefront. In iOS 7, users can change the text size that they see in apps. This dynamic approach to UI requires a different style for app designers.

3 Depth: Motion and layers take the iOS experience to new levels allowing app designers to approach app development in a new way. While iOS 6 saw us often enter apps from the side, through transitions that swept to the left, the current method sees users burrow deeper and deeper into an app, seemingly going down a level. This is achieved via animations, which zoom into the next layer of your selected app.

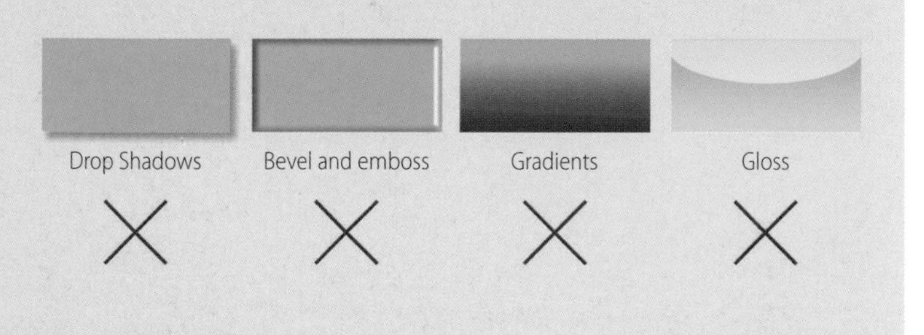

One major design factor to get to grips with is that drop shadows, gradients, and bevels are now a no-no. The clean design of iOS 7 steers clear of making objects look physical. Apps that acknowledge this change will have an edge over those that don't immediately, as these apps will look old-fashioned and out of place in the current environment.

Designers who've been designing for Apple from the beginning will know that innovation, despite the guidelines, has always been central to the best apps. The most successfully designed apps build upon the guidelines and latest platform, innovating on top of what's already there.

Key Design Factors to Consider for iOS 7 and Above

WHAT YOU NEED TO CONSIDER:

You don't have much choice on the points outlined below, though some of the points only apply to those customizing Apple elements.

1 Buttons are borderless: Solid color buttons are out, and transparent, borderless buttons are in. They float on top of the design, so background images need to accommodate this. Most buttons aren't discernible as buttons, they appear as colored text to alert the user that they can press there. Apple suggests using a key color to denote interactivity. Give each interactive element ample room for easy use. Give tappable controls a hit target of about 88 x 88 points.

2 The app icon is bigger: App icons have gotten bigger, and now need to measure 120 x 120 pixels to be approved by Apple.

3 Be retina-ready: If you must support iOS 6 and below, or you're also designing for iPad, then this one's for you. Most designers will have this one down already as the retina screen has been with us for a while. You know the drill, make sure that you create your graphics at double the size they'll need to appear to ensure they don't pixelate on the retina screen. Those designing for iOS 7 and above only need to supply retina screen images. You'll also need to make sure that your app supports iPhone 5's bigger screen.

4 Apps are full-screen: Due to the transparent status bar, apps extend the full length and breadth of the screen. Though you still need to account for the status bar, your background should extend beneath it. This stands for the launch image too.

WHAT YOU SHOULD CONSIDER:

1 Use the whole screen: We're used to seeing insets and frames in many of our apps. Apple suggested that designers rethink this in their approach to iOS 7. Designers are encouraged to use the whole screen in their designs, taking advantage of all of the virtual real estate.

2 Thinner typography: iOS 7 has seen fonts go skinny and lightweight, with Helvetica Regular Neue now being the font of choice. Use iOS 7 system fonts to ensure maximum clarity and improved readability.

3 White space is everything: With function at the forefront, nothing makes the app's primary purpose more noticeable than white space surrounding it. Uncluttered apps with more white space will fit better with the iPhone esthetic.

4 A fine line: Dividing and decorative lines have become thinner, weighing in at just a single pixel in height on the retina screen.

5 Transparent headers: Using the whole screen could only be a possibility if there was a change to the header. Now translucent, the OS allows the background image to continue underneath the header to create a seamless look. Just make sure that you account for the header when arranging your graphics.

6 Translucent navigation: Tab, status, scope, and search bars are all translucent. Make sure your background image accommodates this. The idea is that the blurs beneath the menu or navigational element hint at the activity underneath.

7 The future is layered: To allow this clean and lean interface, something had to give. Gone are the blocky buttons and sliders that obscured the background images; instead they've been replaced by a layered approach which allows the design to shine through. Sliders, buttons, and clocks can now sit on top of your background image due to their transparency; integrate this into your design (see also "The Parallax Effect" opposite).

8 Background images: Your background image has never been more important. The transparent and translucent bars and buttons will make more of your background image visible than ever before. Make it a good one.

9 Remove the splash screen: Though this one applied to the previous OS, the iOS 7 guidelines hammer home the message. As function is the key in iOS 7, Apple want you to be using the app as quickly as possible; the splash screen (the screen you see when first loading an app, often an image) is an unnecessary barrier between you and functionality. Instead, they advise you to supply an image that closely resembles your app when it's up and running; this creates the illusion of speed.

10 Color Palette: Light, bright, and bold, iOS 7 launched with a very contemporary palette. While you are of course at liberty to use whichever colors you like in your app, if you want to fit with the iOS 7 esthetic, you might need to rethink your choices. As stated previously, it's wise to define a color for your buttons (or lack of) which will suggest interaction. Select a palette of colors for your app to give it a uniform feel throughout.

Considerations before you begin your design:

By this stage you should have discussed your navigational options with your developer and have decided on which transitions will power your user's journey. Make sure you factor these decisions into your design.

ALSO REMEMBER:

- **Control Center:**
The Control Center is accessed by swiping up from the bottom of the screen. Try not to interfere with this by placing a similar action in the same place.

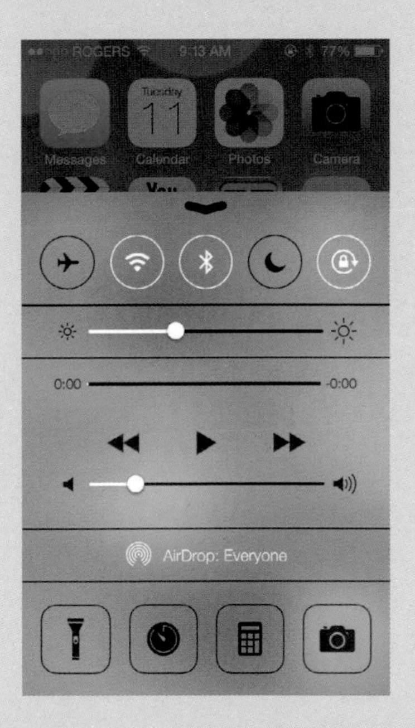

- **The Parallax Effect:**
The Parallax Effect, used to great effect on Apple's home screen, is created via the gyroscope. Whether you're using a panoramic picture or not, the gyroscope will use slight movements to move the different layers that make up a view at different times to create the effect of movement, and to make it seem like the icons are floating. We're guessing that a lot of developers will be quick to try and integrate this effect, but that might cause it to get old fairly fast.

- **Dynamic Type:** It's worth noting that iOS 7 improved accessibility options, which means that users can define the text size throughout the phone. This means that their preference will also cross over into your app. If your app is built using the Dynamic Type function this won't be a problem, as the text spacing, weight, and line height will adapt automatically for the best legibility. It's worth talking to your developer about this.

Dynamic Type
Optical Scaling

Lorem ipsum	Lorem ipsum	Lorem ipsum
Lorem ipsum	Lorem ipsum	Lorem ipsum
Lorem ipsum	Lorem ipsum	Lorem ipsum
Lorem ipsum	Lorem ipsum	Lorem ipsum
Lorem ipsum	Lorem ipsum	Lorem ipsum
Lorem ipsum	Lorem ipsum	Lorem ipsum
Lorem ipsum	Lorem ipsum	Lorem ipsum
Lorem ipsum	Lorem ipsum	Lorem ipsum
Lorem ipsum	Lorem ipsum	Lorem ipsum
Lorem ipsum	Lorem ipsum	Lorem ipsum
Lorem ipsum	Lorem ipsum	Lorem ipsum

- **Fonts:** Though Helvetica Regular Neue is used consistently throughout iOS 7, there are actually 200 system fonts (including their variations). Dynamic Type will support all system fonts, so using one will make your iOS 7 design process a little easier, and make your UX a more pleasing affair due to the benefits Dynamic Type brings.

Key Technical Factors to Consider

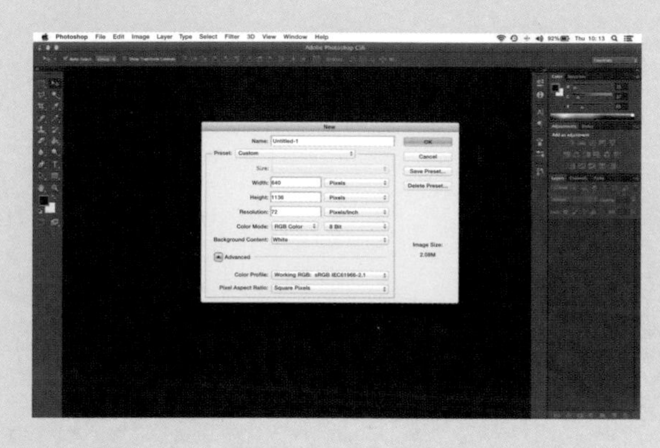

Setting Up Your Photoshop File

While it may seem that we've been talking about designing your app for chapters without actually doing any real design, we've finally reached the stage in which we'll set up the Photoshop file. There are so many different UI components and methods of navigation that it would be silly for us to assume which you'll want to include. Therefore, we're just going to show you how to set up the file to the correct size and settings and add the status bar.

The images and instructions will refer to setting up the template for an iPhone 5.

1 Open a new file in Photoshop, with the following settings: width: 640 pixels and Height: 1136 pixels for iPhone 4, or 640 pixels x 960 pixels for iPhone 4 at 7ppi.

2 Create some native UI areas (by using shapes or simple rulers) according to Apple iOS guidelines, such as:

- **Status bar:** 40 x 640 pixels
- **Navigation bar:** 88 x 640 pixels
- **Tab bar:** 98 x 640 pixels.

(The sizes listed are for a portrait orientation; landscape sizes will obviously differ slightly in length.)

. . . and you're good to go! It looks empty right now, but you should have done enough planning at this point to fill it quickly!

If you don't fancy going through these steps, there are basic templates for each of the iPhones available on our website, www.iphoneappdesignmanual.com, to help you get designing a little faster.

Standard Screen and Icon Sizes

iOS 7 is supported by iPhone 4 and above. There's only been one dimensional change since then in the form of the iPhone 5.

Design iOS 7 app icons at 120 x 120 pixels. Due to a large icon that you'll need to submit for the App Store, which weighs in at 1024 pixels x 1024 pixels, we make our icon at the larger size and scale it down at a later date.

Testing Your Design on a Device

It's best not to take for granted that what appears on your computer screen will look exactly the same on your phone. Not all screens are created equal. Not all retina screens are created equal, for that matter. The iPod Touch features an inferior retina screen to its iPhone counterpart. But what does this mean for your design? Although it's not a huge deal, colors will look duller on lesser devices, and certainly not exactly the same to how they appear on your computer screen.

It's worth checking that your design still looks OK on the handset itself. Our designers use a free tool called Live View. A mobile application design and prototyping tool, it allows us to send screens from Photoshop straight to the mobile device via WI-FI, which is time efficient and very useful. There are many other tools out there that fulfill a similar function.

It's good practice to check your design on every device it will run on. Though it's unlikely that too many problems will arise, it's best to know about any that do before you pass over your app to your developer.

CHAPTER EIGHT

ICON DESIGN

If you're not after something specific, a trip to the App Store for many of us means scrolling through the grid until something catches our eye. It's at that point that you tap through and look at the app in more detail. Only then are you exposed to the marketing text and screenshots which seal the download deal.

The likelihood is that you're flicking past a number of great apps, apps that the designer and developer have spent hours crafting and perfecting, and apps that solve the exact problem that you're looking to solve. But why do they go unnoticed? It's because the icon doesn't appeal to their target audience. A bad icon can kill your app's chances of being downloaded instantly.

The icon acts as a door to the function that lies beneath. If your icon doesn't make the app's function clear, or is poorly designed, the likelihood of someone opening that door to explore the app is slim. In this chapter we'll be looking at the conventions of icon design. By the end, you'll have a clearer idea of how to wrap everything that your app represents into an attractive, one-image package.

iOS 7 Icon Design

Flatter design calls for better design as your image must deliver without the bells and whistles that we've all come to rely on, such as gradients and drop shadows. When iOS 7 changed the way iPhone apps look for the foreseeable future, icons played a major part.

As with the rest of the iOS 7 revamp, the icon too has become flat. Glossy, tangible icons are out, and have been replaced by something a lot bolder. While visual metaphors haven't completely disappeared, they are certainly

not as prominent as they used to be. Gradients do still feature highly, but they're much gentler than they were before. Borders are rare. Drop shadows have pretty much been forgotten. Textures, too, are scarce.

The icons have become rounder, so you'll need to work to the new icon template to get it completely right. While there are only a few pixels difference, the new shape is larger and more distinctive. (NB: Though your app icon will appear to have rounded corners in the App Store, you should design it as a square. The rounded corners are an effect that is applied afterward. By not providing a square icon image, your app would publish strangely, if it was approved at all, which is doubtful.)

Colors are brighter and, perhaps most importantly, the graphics which feature inside them are cleaner and bolder.

The iOS 7 icon template fits with the preference for white space. It encourages you to leave space around the edges to let your graphics breathe.

Removing all of the gloss leaves the design center stage. Fitting in with the design guidelines and succeeding will be testament to your design skills. The old embellishment hid a multitude of sins, a luxury iOS 7 icons don't have.

KEEPING IT IN CONTEXT

Visual metaphors made it easy to keep an app's icon design in context. Icon design for iOS 7 calls for a smarter approach for success. Direct your chain of thought to something more abstract, or if a particular object or shape is relevant to your app's primary purpose, consider incorporating it into your design in an artistically enhanced way. This method can make your app's function clearer to the user. Whatever image you choose to use, just make sure it is relevant and acts as an advert and true representation of what your app does or is about.

The Different Types of App Icons

We can break down the different types of successful app icons into three main groups.

The hyper-real icon: This type of icon features a detailed representation of a real (or fictitious!) object—a cupcake, a fork, a camera. The object is an accurate representation of the app's purpose. These apps replicate characteristics of the object, such as materials used and weight. Though Apple still mention this type of icon in their iOS 7 interface guidelines, they might look a little odd and old-fashioned against Apple's flat icons.

The branded app icon: Quite obviously, these icons feature the logo or easily identifiable brand elements of the company or product. The brand elements are often supplemented by some design feature that denotes the app's primary function. App developers Zynga (see top row icon far right) put their little dog logo in the top left-hand corner of all their app icons.

The preview: These app icons manage to embody the design and fundamental purpose of the app in a single image. This type of app icon is particularly good at reminding the user exactly what the app does. iOS 7 and its veer toward minimalist design has pushed the user to be cleverer with their graphics; rather than using the most obvious image to denote their app's purpose, it's suggested that the designer thinks outside the box.

Our Top Tips for iOS7 Icon Design

1 Use images which are easy to recognize: Though visual metaphors are frowned upon as a whole in iOS 7, the icon calls for universal imagery which is easily recognizable to everyone.

2 Keep it focused: It's advisable that the icon focuses on your app's primary function, definitely not a subsidiary one. Users can feel cheated if they're expecting one thing and get another instead.

3 Build on the design style of your app: This one might sound a little obvious, but there are many apps in the App Store with icons that bear no resemblance to the app's UI design. Don't be one of them! If you've got a particular color palette or style in place for the app, be sure to use it in your icon design.

4 Don't overload it: You might only have a 120-pixel square, but that doesn't mean you should cram it with images. Clarity will allow your app's primary function to take center stage.

5 Don't use text: Your app's name will be written underneath your app. There's absolutely no need to write it on the icon. You'll be wasting space and it will probably be illegible.

6 Don't make the image too complex: What looks good on a monitor might not look as good in a small rounded square on your home screen. You'll lose clarity and detail as it gets smaller, so make sure you can still see what's going on when it's tiny.

7 Never use Apple elements in your icon design: An absolute no-no. Firstly, they're copyrighted. Secondly, it would be confusing to the user to see familiar elements, icons, or hardware in your app.

8 Don't prematurely age your app: By using something like a product replica in your icon design, you're instantly putting a timestamp on it. The minute that product is updated your app will look old.

9 Make it opaque: Don't make your app icon transparent, even if you're trying to mimic an effect like glass. Your app will look like it's floating on a black background and out of place with its surrounding apps.

10 Give it a distinct background: Try and ensure that your background doesn't blend in with what will appear behind it on the home screen. Once people have downloaded your app you want them to be able to see where it is at all times! The transparency of iOS 7 means that when you enter an app's folder, the color of your background will determine the color of the folder. However, folders in the collected view on the home screen have a milky gray background, so it's best to steer clear of a light gray background for your icon. In fact, it's almost always a good idea to steer away from a milky gray background for your app; it's not particularly attention-grabbing.

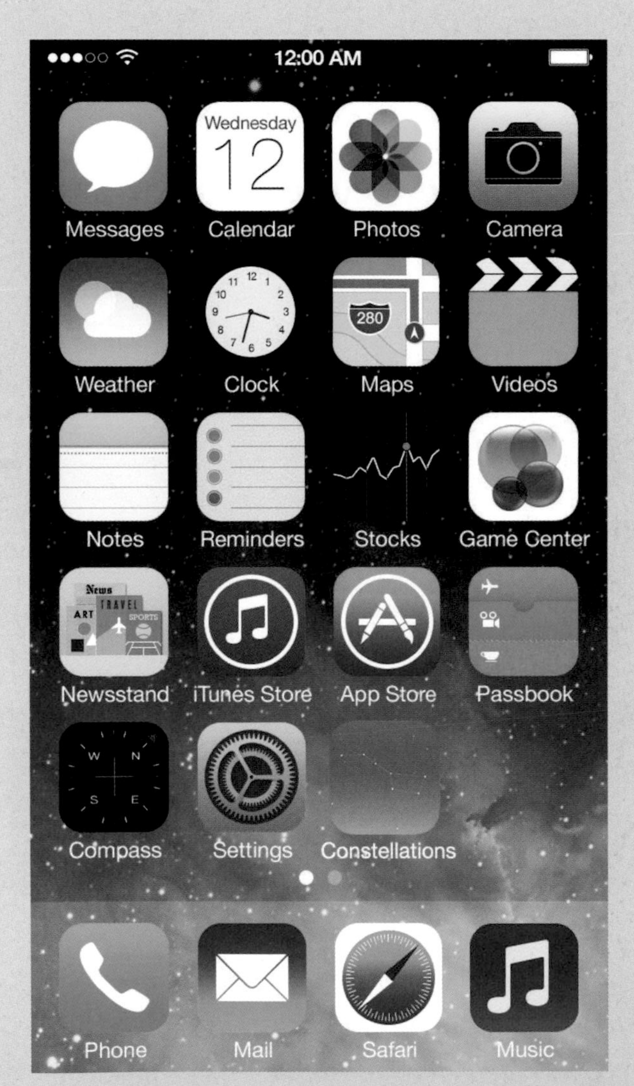

What looks good on a monitor might not look as good in a small rounded square on your home screen.

Icon Sizes and Placements

Your icon needs to work in a variety of sizes and places.

For iOS 7, you'll need to provide an icon at 120 x 120 pixels. As iOS 7 isn't supported by devices with a non-retina screen, you don't need to worry about providing the images in non-retina resolution.

You'll need to supply a large app icon for use in the App Store at 1024 x 1024 pixels. Design your icon at 1024 x 1024 and then size it down, checking that all relevant detail still works at the smaller size.

You'll also need to supply a small icon that will be used in the settings center. This will need to be 58 x 58 pixels. Make sure that your icon translates and is easily recognizable in such a small state. People will need to pick your app out of a list of equally tiny icons; make sure yours stands out.

It's recommended that you supply an icon for Spotlight Search at 80 x 80 pixels. There are standard conventions that you should conform to when naming your icon and its size variations. There's a full table of naming convetions at the back of the book. These icon convetions are:

Icon@2x.png The app icon for the retina screen (iOS 7).

Icon-small@2x.png The name for the search results icon on iPhone and iPod touch. This file is also used for the settings icon on all devices.

People will need to pick your app out of a list of tiny icons; make sure yours stands out.

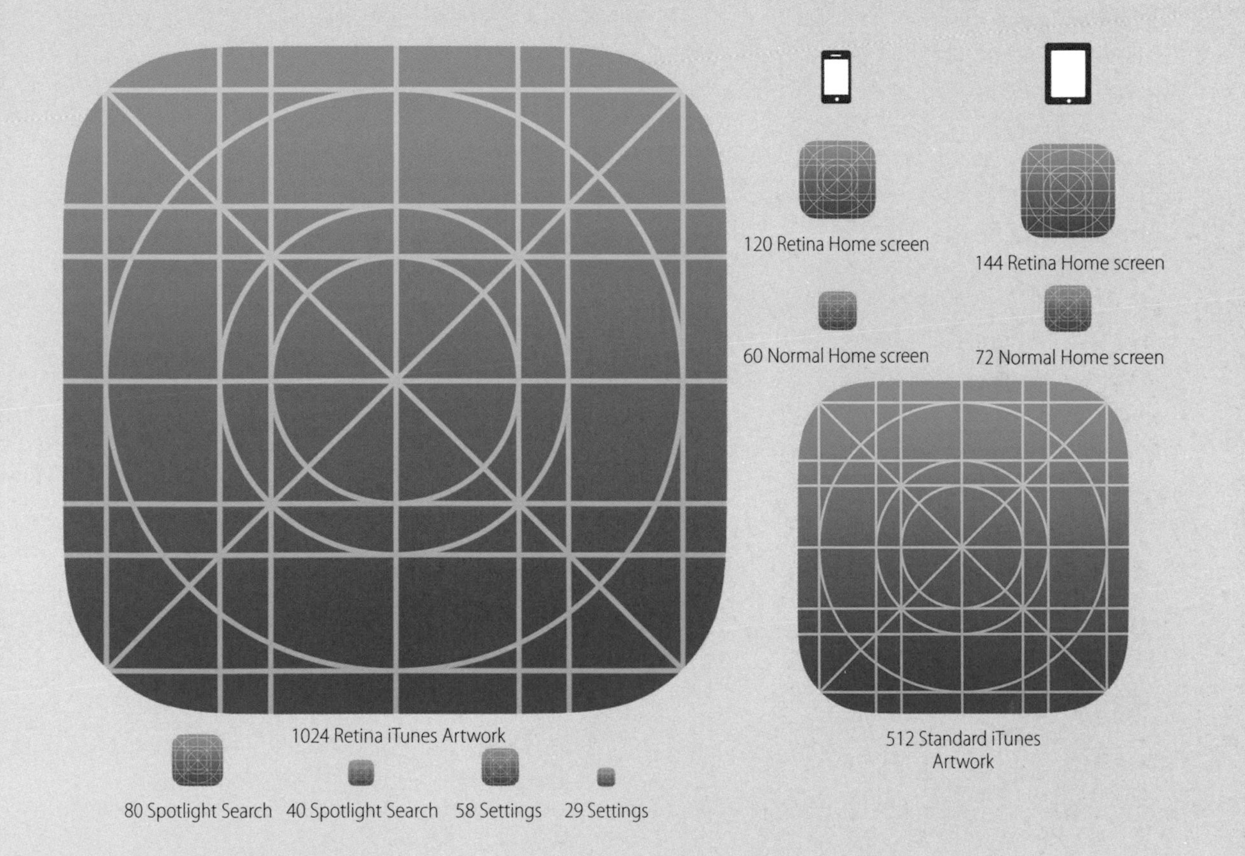
120 Retina Home screen
144 Retina Home screen
60 Normal Home screen
72 Normal Home screen
1024 Retina iTunes Artwork
512 Standard iTunes Artwork
80 Spotlight Search
40 Spotlight Search
58 Settings
29 Settings

Key Factors to Consider for Icon Design

Even if you've followed all of the suggested guidelines for icon design, there are still a few things that we've come across that can make or break an icon. The most visually appealing icons can fade into obscurity if they don't follow a few simple rules. Luckily for you, we've outlined them here . . .

Graphical Continuity

As we've stated, graphical continuity between your UI, icon, and loading image is a must, even more so with iOS 7. One method of achieving this is to use elements of your UI in your icon design. Some successful icon designs have cleverly wrapped their UI into their icon. This makes it terrifically easy for the user to remember what your app does and to pick it out in a line-up.

Obviously, if your app has a strong and well-recognized logo, it makes sense to use that in the icon. Colors should match or complement the colors found in your app.

Remember that you're aiming for a consistent experience. Using the same style and colors for all three elements will give the impression of a coherent and professional application. An icon design that fits with the load screen image will add to the impression of a speedier load time.

Test Your Icon Design on a Device

There are plenty of programs that will allow you to test your icon on your phone. It's important to see if your app icon can hold its own against the others out there. There are many free and easily downloadable tools available that will allow you to test your app on your home screen, such as Skala Preview's Skala View and Flinto's Icon Strike.

As with your UI design, different devices will display your icon differently due to factors such as the retina display. Make sure that you're happy with the final result on all of them, but always remember to look forward. People upgrade, and upgrading is good! You want your design to look as good as it can for the longest time possible. Though we still support older models with our design, we ensure that our design looks its best on the iPhone 5.

Capitalize on Your Brand

Some well-known brands started life as an app. Take Instagram as an example. The original Instagram App Store icon is now instantaneously recognizable to web and mobile users across the globe. Like all good logos, the icon acts as a conduit to the brand and embodies what it's about and what it stands for.

As we've said before, even if your app is a stand-alone product with no plans for expansion or for use on different platforms, thinking of it as a brand in its own right wouldn't be a terrible thing. Brands have their own set of guidelines, which set out the fonts, colors, and logo recommendations that all of their communications and platforms will feature. By making these rules for your app you'll keep the design cohesive, and these guidelines will without doubt influence your icon design. If you want to expand your app to other platforms then you need to take this into consideration when designing your icon. You'll need to pick something that's transferable and scalable.

Those of you designing for a business with its own distinctive brand should capitalize on this. Your brand guidelines will give you great advice on which colors to use and how to use the logo.

If your logo is recognizable enough on its own, it's definitely not a sin to use it as the app logo. It's worth considering whether this will be the only app that the company you're designing for will publish in the App Store. If it's one of many, then using the logo will be out, as the app will need to be more distinctive and portray or hint at the function that this app fulfills. Don't just take for granted that your logo will work as an icon as is. It still may need to be tweaked to stand out, and it will still need to adhere to the conventions and advice that we've outlined. As the App Store icon is relatively small, using a large image which works on the web may end up looking like a dirty smudge.

Standing Out in the Grid

There's no way of knowing what your icon will be surrounded by in the App Store or on a user's home screen. You need to give your icon the best fighting chance that it can have. Though you may be settled on an icon design, you might want to consider A/B testing to ensure that you're going down the right path and that your icon resonates with your target audience. Even color variants can have a massive impact on how your app is perceived and how much it stands out to your target audience.

When you've settled on an icon design, it's worth creating a few variants. There are many ways that you can test out the effectiveness. Using a screenshot of the App Store grid, paste one of your icon variants in there and test it on family and friends against another of your designs. Use social media channels, and trusted forums. Get opinions from everyone you possibly can. By giving people options, they're more likely to be forthright with their opinions on what's wrong and right with your app.

CHAPTER NINE

SCREENSHOT DESIGN

Your app might be the best thing on the market but there are a multitude of things that can prevent people from discovering it. Before you submit your app to the App Store, we'd advise you to look into discoverability and app marketing.

Keywords, your app's name, pricing, and the marketing text are just some of the considerations you should bear in mind. While developers have yet to crack it, you should be aware of best practices to optimize your visibility through the App Store's search algorithm. This factors in your app's name, keywords, ratings, and popularity to rank the search results that appear to people. We're not going to be exploring these areas in this book, just the design, but we've included a library of resources that can start you off on app marketing. It's a volatile market out there. Ensure your beautifully designed app stands a good chance of being noticed.

One marketing element that we can discuss are the screenshots. Much like an advert, the screenshots act as the packaging for your app. They give the user a taste of what they can expect, a slice of your UI, and an idea of your app's quality and usability. This chapter will set about explaining how to choose which screenshots to use (and if they need to be actual screenshots at all) by looking deeper into the app-buying process and the psychology of the buyer.

How We Decide Which Apps to Buy

It's really quite simple. People want quality apps that fulfill the function they claim to, and look good to boot. After spending a copious amount of time working on their app's UX and UI, many designers appear to fall short on this selling point through their choice of screenshots.

But what is it that people are looking for? When we enter a search term in the App Store, we're usually presented with a huge amount of apps to choose from, unless, of course, our criteria is very specific. What makes us pick one app over another?

There are a number of things that people look at when choosing which apps to explore further:

1 The name: Potential customers tend to look for apps which have a name closely related to their search query. While you don't have to explicitly say what the app does in the title, some connection is better than none.

2 Ratings: Apple displays the number of stars and reviews that an app has on the search results screen. The more stars and amount of ratings, the better, and the more chance of someone delving a little deeper.

3 The price: Can they get the same thing for cheaper or for free? Despite most paid-for apps being the same price as a chocolate bar, people are reluctant to waste their money. If a free app performs the same function as one that's $2.99, which do you think they'd choose?

4 Screenshots: The main screenshot you choose is displayed in the search results under your app name. This is one of the most important factors for the majority of buyers. The rest only get a look in if the first one does its job and sparks interest.

Potential customers tend to look for apps that have a name closely related to their search query.

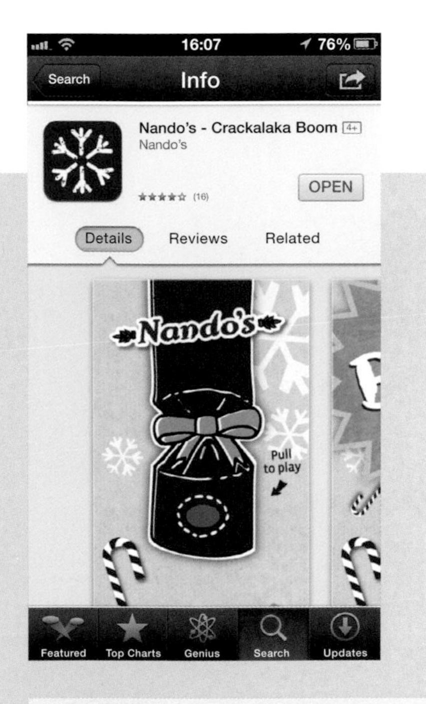

You click on the icons of apps with well-designed screenshots that have scored well in reviews, and appear to offer you what you need at the best price. But how do you ensure that your app is one of them? When your app is first published, it will have few reviews, so it's the job of the screenshot and logo to entice your potential customers.

Many people don't bother to read the marketing text until they're just about to buy; they won't even see it if they don't tap through from the search screen. The lesson learned here is not to leave vital information out of the images. Don't rely on your text to inform the potential customer of functionality; the text won't get a look in if your screenshots aren't doing it for them.

Selecting Your Screenshots

Your main screenshot needs to relay your whole UX in a single frame—a daunting task to say the least. Many designers forget about the screenshots until just before they submit, quickly bringing together a few shots of the UI.

If your app is a single-screen app with relatively no action, then chances are this will be good enough. But the more complex your app idea, the more information you need to convey in just one picture. Though you can submit up to five screenshots to the App Store, it's generally the first one that people use to decide whether to look further at your app. The other four are there to reaffirm their faith in the first and to show off alternate functions that your app features.

Like with any product design or logo, you're never going to please everyone with your screenshots, but there are a number of things that you can do to ensure that you make the best impression possible.

Firstly, your screenshots don't have to be actual grabs of the app in action. You can create them in Photoshop for greater control over the clarity. You can make tiny tweaks to make an action or element clearer. We don't recommend embellishing on the actual design—it would be a bit like embellishing on the truth. People like to know what they're getting. If your shots differ greatly from your app, expect reviews to reflect this. That being said, there are a few flourishes which you can add that will enhance the first impressions people form of your app. If your app features a lot of interactivity, sounds, or moving images and animations, these factors might not convey in a still image. Feel free to add subtle shapes and icons which demonstrate these elements. Do you need to swipe to action your app? An arrow across the screen could simulate this action to the viewer. This type of screenshot adornment also serves to educate the potential user on how to use the app. This makes the app seem more user-friendly from the offset.

Your screenshots don't have to feature an image of your app in every picture. Many people use the space to showcase something different. Some people use the pixels for an advert-style image that sells the benefits of the app. Some people use their logo (use with caution if your logo isn't well known).

One device that appeals to many App Store users is seeing the app in action, being used in someone's hand and fulfilling its function before your very eyes. Apps such as Clear and Vine do this well. The user can see immediately what the app does and the type of results you can expect to get from it. Consider using this method if your app does something unexpected or new-fangled. It will add to the innate usability should they choose to download it.

Don't think that text is a no-no; if a few words added to an image add value, go for it. Some developers choose to leave instructional messages in their screenshots, increasing the user's knowledge of how to use it pre-download.

Another device employed by app designers is to extend an image across multiple screenshots. It was used to great effect with Bump. The way that the App Store is designed allows the screenshots to sit side by side, with users scrolling their way through them. You can take advantage of this by extending your image across more than one, slicing the image up into two to five sections. This is particularly useful if it's difficult to fit the main body of action in just one. A word of caution with this approach: if you choose to create something along these lines, make sure that your primary image is good enough when it appears on its own. There's only one screenshot shown on the first search screen. Don't squander it by trying to be too clever. Perhaps consider one standalone image and linking a few of the others instead. Another consideration for this method is the possibility of it being reviewed. If your app is picked up by journalists, many take screenshots from the App Store for their reviews. If your images are all part of one continuous image, they will be useless in this context.

Like any great ad, your screenshots should tell a story, namely the story of your app. The messages should be clear. They should be succinct. They should be direct. They should sell your service and UX. Oh, and they need to look amazing.

Our Top Tips for Screenshot Success

1 Have a well-designed app: This one sounds a little silly, especially in the context of this book, but it's true. If your app is a delight to behold, it stands to reason that the screenshots will be visually pleasing. Make each of your screens a joy to look at so that there's not a single screenshot that lets the rest down.

2 Have continuity between your app design and your screenshots: Again, this might sound stupid, but if you've decided to go down the route of using alternative images, make sure that they still give a flavor of what's to come, not something completely different. Use the same or a complementary color palette to the one used in your app. The same goes for fonts.

3 Show what your app actually does: Viewers will thank you for it. Nobody wants to waste their time and bandwidth on the wrong thing. Make it easy for people with screenshots that demonstrate your product perfectly.

4 Add text and icons where necessary: Don't be afraid to add a little textual embellishment or an icon here and there if it helps you to better demonstrate your app's function.

5 Don't include the status bar: Despite iOS 7's move to full-screen apps which feature the status bar as an overlay, Apple still advise that the status bar should not appear in the screenshots. Some apps, gaming apps in particular, rarely show the status bar anyway, so their screenshots are uploaded at full-screen size. If you do show the status bar in your app, you'll need to crop the status bar out of the screenshot and submit the image at the smaller size. (Screenshot sizes are listed opposite.)

6 Only use your icon as your main screenshot if your logo is well-known: Though you might be proud of your logo, if it doesn't mean anything or add some level of understanding to what your app does, save it for one of the secondary ones if you must use it at all.

7 Localize: If you're going global with your app, don't forget to extend the courtesy to your screenshots! It's annoying to see text in a different language, especially if your language skills aren't great. (Top tip: don't rely on Google Translate to get your message across. The translations aren't always grammatically correct. If you can't afford a translation, see if any of your friends speak another language.)

Ultimately, the strength of your UI will do a lot of the talking. Make sure you test your images with people you haven't explained your app to. See if they can gauge what your app is about from just the name and screenshots.

Screenshot Specifications

As mentioned, the first screenshot uploaded will be the main one that appears in the scrolling search results bar in the App Store on your phone. It will also be the first to appear in iTunes and App Store product pages. The order of upload is therefore important. If you're not submitting them yourself, make sure that whoever is knows the order that you want them to appear in.

Make sure the first one is the most striking and the one that says the most about your app. You're allowed four more on top of this; they will also appear in the order in which they were uploaded.

All screenshots need to be 72ppi and in RGB. Acceptable file formats are JPEG, TIFF, or PNG. No other formats are accepted, nor is it possible to submit zipped files. There should be no transparency in your screenshots.

You will need to supply images for the 3.5-inch display (iPhone 4s and below), and the 4-inch (iPhone 5 and iPhone 5s). If your app will be available for iPad, you'll also need to supply shots sized for this.

If you're using the maximum number of screenshots (we advise this), you'll need to submit ten images. You need to really give these some consideration, as if you want to change them, you'll have to submit an update to Apple. This can take up to three weeks (more in rare cases) to be actioned if it's approved. Three weeks is a long time in the app world. Keep this in mind before you submit.

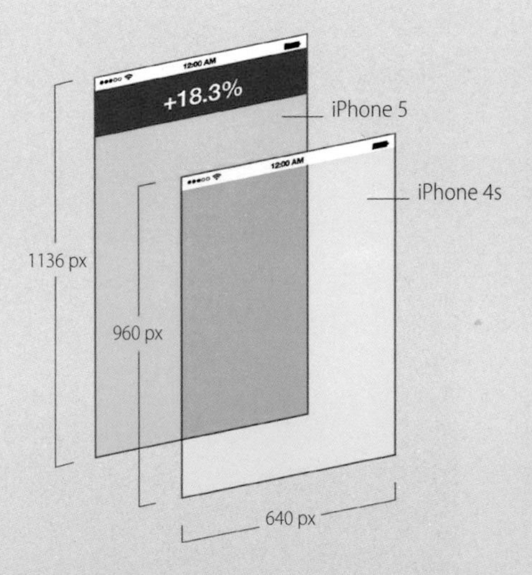

SIZES FOR THE 3.5-INCH SCREEN:

Portrait:
Minimum requirement:
640 x 920 pixels
(without status bar)
Maximum requirement:
640 x 960 pixels (full screen)

Landscape:
Minimum requirement:
960 x 600 pixels
(without status bar)
Maximum requirement:
960 x 640 pixels (full screen)

SIZES FOR THE 4-INCH SCREEN:

Portrait:
Minimum requirement:
640 x 1096 pixels
(without status bar)
Maximum requirement:
640 x 1136 pixels (full screen)

Landscape:
Minimum requirement:
1136 x 600 pixels
(without status bar)
Maximum requirement:
1136 x 640 pixels (full screen)

CHAPTER TEN

PREPARING FINAL ARTWORK

When your design is complete, you may be under the impression that the hard work is done and that you can hand over your Photoshop file as it stands to the developer. This is not always so. Your programmer will have enough work to do, and presenting him or her with another screen full of work might not endear you to them. There are some developers who will take on this task for you, if you ask nicely!

You will need to have each of your elements ready in the required sizes for the retina and non-retina screens. You also need to adhere to some naming conventions, which we'll go through later in the chapter.

All of your files will be saved as Portable Network Graphics (PNG) files. This is standard across iOS native apps due to the way that Xcode, and later your app, will optimize the image. You can use JPEGs, but try to only use them where absolutely necessary.

Slicing Graphics

Slicing is the term that we give to taking your completed 2D design and "slicing" it up into the individual components, or digital assets, which make up your app. From the back button to each of your tab icons, they all need to be saved separately.

If you have included drop-shadows, gradients, and the like to create the look and feel that you want, it's not just a matter of exporting each layer individually. You need to hand over the finished article to your developer so that the image appears exactly to your specification.

There are many ways that you can do this, but we'll describe the method favored by Apposing's designers; as we use Photoshop, our instructions are based on using this software. Firstly, you need to identify the different layers that make up your component. As all Photoshop (or PSD) images are organized in layers, you can isolate each element from the others.

You need to identify the layers that feature your component, and then use the Eye Icon beside each of the layers not included in the component to switch off their visibility. You can do this quickly by holding down ALT while clicking on the Eye Icon of one of the included layers. If your component features only one layer, then you're done for this step. If your component features multiple layers, turn the visibility back on for each of these.

You should be left with just your component visible on the screen, whether it's made of multiple layers or just one. Locate the Trim tool in the Image menu. This allows you to remove all of the transparent pixels and leaves the pixel shadows. You could use the Crop tool here, but it's not as efficient. After you've actioned Trim, you should just be left with your component with minimal or no transparent pixels.

The Photoshop plugin we mentioned earlier, DEVROCKET, can save a lot of time when exporting.

Now it's just a matter of saving your image as a PNG file. You need to do this with each individual component in your app. There are naming conventions for specific components, which we get to in the next section.

Other methods include selecting your layers in the same way and converting them into a smart object, then saving them as a PNG on the next screen. There are also tools out there, such as Slicey, which allow you to define parameters (either by naming layers or boundaries) to export images. Software such as this also allows you to save images in both of the required sizes.

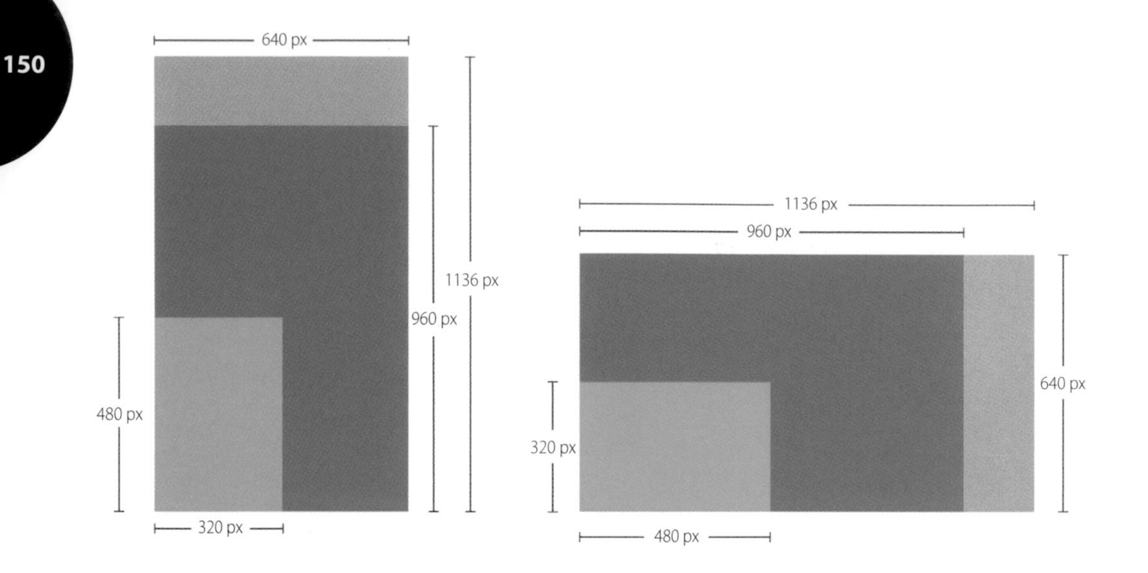

As we discussed earlier, if you'll be supporting iOS 6 or if you're designing for iPad, you'll need to supply your digital assets in two sizes for the retina screen. If you've done as we suggested and designed for the retina display, you'll need to reduce each component by 50% and re-save. The first image you save should be named yourfilename@2x.png and when reduced by half it should be saved as yourfilename.png. If you're making an app for both iPhone and iPad, you'll need to add the suffix "~iphone" or "~ipad" (all lowercase) to the retina filenames. If you're lucky enough to have a developer who will export your design for you, firstly make sure they're comfortable with the software you've used. The industry standard is Photoshop. You'll need to send them the PSD file as is, so it's vital that your layer naming structure is in place. Name layers logically. Make sure every component is saved in a separate layer. Also make sure each layer is named something easily identifiable.

NAMING CONVENTIONS

You'll no doubt have triple-figure numbers of components to export. If not, count your lucky stars! Naming them in a logical way can be of great benefit to your developer. Again, it's worth talking this through before you start exporting to save you time. It's possible that your programmer already has a series of naming conventions in place. For instance, they might prefer for images that appear in the tab bar to contain the word "tab."

There are a number of images, such as the launch image, which have definitive naming conventions. The launch image is always named "default." So after you've exported your launch image in the two required sizes, you should be left with Default.png and Default@2x.png.

Preparing Files for the Developer

The way that you organize the saved images, and, in turn, the sounds, is down to preference. If you're not programming the app yourself, you might want to ask your developer what file structure is best suited to them. Organizing your images and sound files in the way they desire will make the app development process easier, and could also save time. As your developer's time is most likely expensive, this step is worth it.

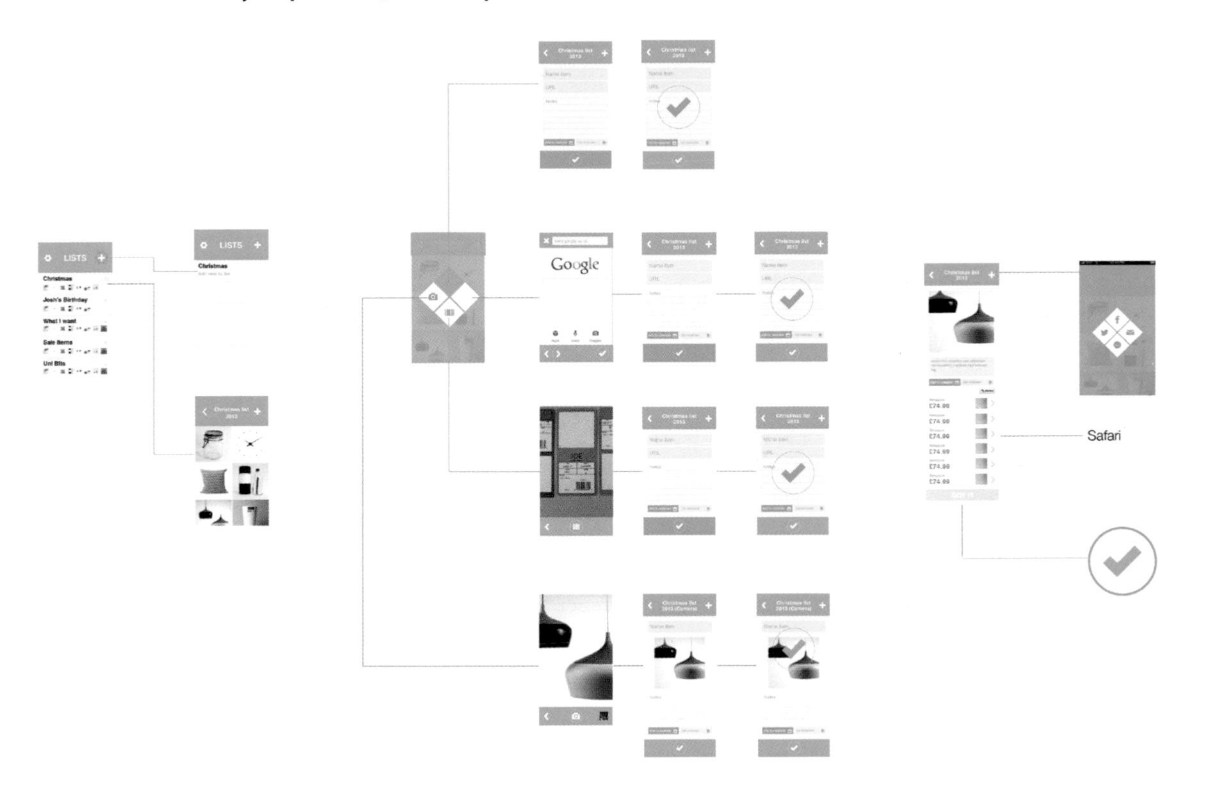

It's also worth creating an annotated document which features every screen of your app that they can use for reference. This should feature everything, from the font used and line spacing throughout to notes on navigation styles and animations. This leaves less room for error, and the developer can refer to it for the majority of questions that might crop up. There are also a number of screen casting tools which allow you to record a short video to talk about each screen.

CHECKLIST

We've compiled a checklist to go through before you hand over your baby. You should have already considered the majority of these, but going through them can save red faces should you have forgotten something . . .

1 Does your app icon show the function of the app in a glance, and is it easily identifiable?

2 Is there an apparent uniformity and consistency between all screens, screenshots, and the icon?

3 Have you remembered to include the launch screen and icons in the required sizes when packaging up your assets?

4 Are all tappable areas at least 88 x 88 pixels (actual size)?

5 Does your app feature buttons or features that become active or disabled? Make sure you have an image that corresponds to each of these states.

6 Are all dimensions of design elements even? Remember, there are no odd numbers allowed.

7 Are all your PSD layers or images named logically? (Also try to stick to the file organization structure recommended by your designer.)

8 We're sure you've got this one covered, but do all of your buttons lead somewhere or do something?

9 Have you included a cheat sheet for your app, containing the final design for each screen and anything else the developer needs to know?

If you can answer yes to all these questions, and above all you're happy with the UI and UX, then you're good to go. Your developer is likely to have questions, so be available to answer them. Good developers will come to you with problems and offer solutions. The solution might require more work on your end. The developer might also come across issues that you hadn't foreseen. Don't be unreasonable in this instance. While programmers can certainly do magical things, they're not actually magicians. Sometimes there's a workaround to a problem, sometimes there isn't. Keep this in mind throughout the process.

CONCLUSION

Hopefully what has become clear is that app design and development is a very personal thing. Each well-designed app is a reflection of the way its designer views the world, and shows a way to solve a problem that he or she has come across. Though we could have given you a step-by-step guide to designing one of our apps, we don't think that would have given you the tools to make your own ideas shine. It would have taught you how to design like us.

Instead, we've tried to give you the information we think you'll need to know to design your first app in the way that suits it best. We've tried to pose the questions we wish someone had asked us before we sailed into uncharted territory all those years ago.

Another important thing to realize is that we're still learning. With every update and every new model of iPhone, we're given the opportunity to create new ways to communicate our ideas and to interact with people. iOS 7 changed the way we approach app development infinitely. The day you think you know all there is to know is the day that your apps will be left behind. To keep up, you need to read news and forums. We gorge ourselves on clever apps that others have created, of stunning examples of custom UI, and clever navigational methods. We applaud other developers' creativity and invention, and learn from it. No app designer is an island; we feed off other to create our best work.

iPhone interactivity has come on in leaps and bounds since we started out in the industry, and it will only continue to improve. Embrace the future; wrap what you learn around your ideas and integrate the platform's best practices. Make mistakes, take chances, and then learn from them. Never forget the lessons learned from previous ventures and don't let failures hold you back. Very few of the world's leading app development studios had resounding success with their first title, but each mistake leads to a stronger next try.

We've added templates, resources we find useful, and more to
www.iphoneappdesignmanual.com
We can't wait to see what you come up with. Share your finished apps with us at:
@iphoneappdesign
We leave you with our company mottos, which we try to integrate into every facet of our working life: The first is, "The best way to predict the future is to invent it," and the second, "Simplicity is the key to brilliance."

GOOD LUCK!

All at Apposing

www.apposing.co.uk hello@apposing.co.uk

Glossary

3G: Third-generation wireless mobile telephone communications. Applications include wireless internet calls, mobile internet access, fixed wireless internet access, video calls, and mobile TV.

4G: Fourth-generation wireless mobile telephone communications. Applications include improved mobile web access, gaming services, high-definition mobile TV, video conferencing, and cloud computing.

Accelerometer: An accelerometer is a sensor which measures the tilting motion and orientation of a mobile device.

API (application programming interface): A specific method approved by a computer operating system or by an application program which allows programmers to make requests of the operating system and hardware, or another application, when building software; i.e. recent Apple APIs allow developers to make use of iPhone functions such as the compass.

App Store: Online store for mobile device users. iOS developers submit applications to the Apple App Store, which, when approved, are able to be purchased and/or downloaded by consumers.

Apple Developer Portal: A range of resources for developing, designing, and distributing applications for iOS.

Augmented Reality: A type of virtual reality, augmented reality software generally creates a composite or alternative view that fuses elements of the real world with elements generated by a computer.

Badge: A visual notification bubble that alerts the user to information such as a new email.

Bluetooth: A wireless technology which allows the user to exchange and broadcast data over short distances from fixed and mobile devices.

Best practice: A tried and tested process or technique that produces required results.

Bug: A coding error in a computer program.

Build: A version of a piece of software; or as a verb, "to build" can mean either to write code or to put individual coded components of a program together.

Development Process: A set of tasks performed for a given purpose in a software development project.

Device Orientation: A term that refers to whether the phone is held in landscape or portrait mode.

End User: The group of people that a software program or hardware device is intended for, or to be used by.

File Transfer Protocol (FTP): A standard network protocol commonly used to transfer files to or from a server over the internet.

Functionality: Refers to the capabilities or usefulness of a software program or application.

Global Positioning System (GPS): A hardware function which provides location (latitude and longitude) and time information by using satellites around the globe.

Graphical User Interface (GUI): A user interface in which the user interacts with the device via images instead of text commands.

Gyroscope: A sensor found in mobile devices which measures the angular rotational velocity.

Hardware: A mobile device of any kind or additional components.

Hit Target: Touch points on touch-screen devices. The smallest recommended tappable area is 88 x 88 pixels (or 44 x 44 pixels on a non-retina screen).

Hybrid App: Ultimately, web apps with a native outer shell or wrapper. Hybrid apps are a software bundle that is a combination of a native app (uses programming language Objective-C for iOS and Java for Android) and a web app (HTML5, CSS, JavaScript). Hybrid apps can use native device functionalities (such as the camera, GPS, Bluetooth, and accelerometer) and be distributed as native applications in the App Store.

In-App Purchase: Users are able to purchase new content, services, complementary features, subscriptions, and upgrades within a free or paid-for mobile application.

Interaction Design (IxD): The practice of designing interactive digital products, environments, systems, and services, tailoring the interaction for an easy and productive experience for the user. The best interaction design occurs when human behavior is observed and the learning applied to ergonomics, gestures, transitions, and mobile-specific interaction patterns.

Interface Design: See User Interface or UI.

Jailbreak: The illegal action of removing the limitations imposed by Apple on iOS devices. "Jailbroken" iPhones and iPads can download unapproved and pirated apps.

Mobile App/Application: Software created for use on a mobile device, which is either pre-installed by the manufacturer or downloaded and installed by a consumer.

Mobile Browser: A web browser designed to be accessed by mobile devices and optimized to display web content effectively for small screens.

Mobile Carrier: Also known or referred to as wireless carrier or wireless service provider; a telephone company that provides services for mobile phone subscribers (Verizon, AT&T, EE, Vodafone).

Mobile OS (Operating System): An operating system for mobile devices, similar to Windows OS and Mac OS for desktops and laptops.

Native Application: A software bundle developed and designed using the programming language for that specific platform (Objective-C for iOS, Java for Android), which runs only on specific devices.

Programming Language: An artificial language used by computers to control the behavior of a particular piece of hardware. Each language has its own nuances. In terms of mobile, iOS apps are written using the programming language Objective-C. Android apps are written using the language Java. Hybrid apps are written using multiple languages via third-party software.

Push Notification: Notifications received from servers on an iOS device which can be viewed without opening up the application.

Retina Display: Apple's liquid crystal display screens found on iPhone 4 and above, and third-generation iPads and above. The number of pixels per inch on these screens is touted as "retina quality," meaning the user's eyes will not be able to spot any pixelation at regular viewing distance. Designers needed to provide images at double the size to accommodate the retina screen due to the increase in pixels per inch after the retina screen was introduced.

Screenshot: An image taken by a device depicting what is being displayed on the device's screen at that moment. Also referred to as screen grab or screen capture.

Software Development Kit (SDK): Development toolkits allow developers to create applications for a particular software platform or framework. Apple's SDK features Xcode, an integrated development environment (IDE), or the piece of software which developers write their code in to create the app, as well as a raft of technical resources and tools.

Touch User Interface Gestures: These are gestures (such as pinch, tap, zoom, and swipe) which are used to interact with Apple, Android, Blackberry, and Windows phone devices. The gestures have different meanings across different platforms, although many similarities occur.

Touchscreen: Standard across most smartphones, touchscreen is an electronic visual display that detects the presence and location of touch within the display area, allowing the user to interact directly with a device without having to use external hardware such as a mouse.

Use Case: A methodology that allows software designers to determine which features a piece of software will require. This is achieved by creating a set of user circumstances, goals, and interactions which serve to alert the developer as to whether these goals can be achieved in the proposed environment and any possible problems which may arise.

User Experience (UX): Indicative of how easy to use and intuitive a piece of software is, the user experience refers to how an end user felt or reacted after interacting with a product. The navigation, user interface, and interaction design all contribute to the user experience.

User Interface (UI): The visual face of your app, designed with the end-user's experience in mind. The UI must present the app's functions in a manner that makes using it easy to understand, as apps don't come with a manual. The iPhone app UI must conform to certain Apple-defined criteria to keep the app experience uniform across the platform.

Wireless Application Protocol (WAP): A set of protocols that allow mobile devices to connect to the internet via a WI-FI connection.

WI-FI or Wireless Network: A method for connecting devices to the internet using radio waves instead of a wire. A device's wireless adapter transmits, receives, and decodes data moving between the device and a wireless router.

Wireless Services: Mobile telephone communications such as 3G or 4G.

References

Apple Human Interface Guidelines: Apple (2013). iOS Developer Library: https://developer.apple.com/library/ios/design/

UIKit User Interface Catalog: Apple (2013). iOS Developer Library : https://developer.apple.com/library/ios/design/

Websites for prototyping and wireframing

iPhone App Design Manual—for all of the resources we mention in the book: http://www.iphoneappdesignmanual.com/

Balsamiq—wireframing software: http://balsamiq.com/

Moqups—wireframe and UI concept tool: https://moqups.com/

Pop—prototyping and wireframing tool. http://popapp.in/

Skala Preview—view app interface and icon designs on your phone with this: http://bjango.com/mac/skalapreview/

Market Research

App Annie—App Store statistics and analytics: http://www.appannie.com/

XYO.net—App Store statistics and estimated sales figures: http://xyo.net

Marketing

Flurry: http://www.flurry.com/

Google Analytics: http://www.google.com/analytics/

Tokens—track the promo codes that you send to friends, influencers, and journalists: http://usetokens.com/

Apptamin Cheat Sheet: http://www.apptamin.com/blog/app-developer-aso-cheat-sheet/

IU ELEMENT	WIDTH	HEIGHT
Status Bar	320px	20px
Navigation Bar	320px	64px
Text Field	variable	30px
Segmented Control	variable	29px
Slider	variable	34px
Switch	51px	31px
Activity Indicator	20px	20px
Progress View	variable	2px
Page Control	variable	37px
Stepper	94px	29px
Picker	variable	162px
Search Bar	variable	44px
Toolbar	variable	44px
Tab Bar	variable	49px
Keyboard (Portrait)	320px	216px
Keyboard (Landscape)	320px	162px

FILENAME	SIZE	USED FOR
Icon-60@2x.png	120x120px	App Icon
Icon-40@2x.png	80x80px	Spotlight Icon
Icon-small@2x.png	58x58px	Setting Icon
iTunesArtwork@2x.png	1024x1024px	App Store Icon
Default@2x.png	640x960px	iPhone 4s, 4
Default-568h@2x.png	640x1136px	iPhone 5s, 5c, 5

DEVICE	WIDTH	HEIGHT
iPhone 5s, 5c, 5	640px	1136px
iPhone 4s, 4	640px	960px

Index

Index

Acknowledgments

I'd like to thanks my partner Becky, daughter Eleanor, and all my amazing family and friends. Massive thanks to Vicky and my team at Apposing, without whom this book would have not been possible.

Thanks to our editor, Ellie for her patience, and everyone at Ilex Press.

Special thanks to Think Moto (http://www.thinkmoto.de/) for the use of their icons on page 71. Playrise Digital, Mailbox, Dropbox, Amazon, Vine, Spotify, YouTube, Flipboard, Spotify, Facebook, Twitter, Halfbrick Studios, Vestaz, Mule, Algoriddm, Us Two, Rovio, Tap Tap Tap, Tapbot, Egos Ventures, National Geographic, Pop, Omnigraffle, Balsamiq, Moqups, Codegoo, eBay, Realmac Software, Bump for allowing us to use their screenshots and icons.